# CONVERGING
# REALITIES

# CONVERGING REALITIES

## TOWARD A COMMON PHILOSOPHY OF PHYSICS AND MATHEMATICS

## Roland Omnès

PRINCETON UNIVERSITY PRESS

PRINCETON AND OXFORD

First published in France under the title *Alors l'un devint deux*
(*La question du réalisme en physique et en philosophie des mathématiques*)
© Flammarion, Paris, 2002 English translation © 2005 by Princeton University Press
Published by Princeton University Press, 41 William Street,
Princeton, New Jersey 08540

In the United Kingdom: Princeton University Press, 3 Market Place, Woodstock,
Oxfordshire OX20 1SY
All Rights Reserved

Library of Congress Cataloging-in-Publication Data
Omnès, Roland.
  [Alors l'un devint deux. English]
  Converging realities : toward a common philosophy of physics and mathematics /
    Roland Omnès.
      p. cm.
  Includes bibliographical references and index.
  ISBN 0-691-11530-3 (acid-free paper)
    1. Mathematics—Philosophy.   2. Physics—Philosophy.   I. Title.

QA8.4.O4513 2005
510'.1—dc22                                                      2004046641

British Library Cataloging-in-Publication Data is available

This book has been composed in Palatino and Swiss 911.

Printed on acid-free paper. ∞

pup.princeton.edu

Printed in the United States of America

10 9 8 7 6 5 4 3 2 1

# Contents

PREFACE   vii

*PART ONE  -  PRELIMINARIES*

**CHAPTER ONE**
Beginnings   3

**CHAPTER TWO**
Brain and Reality   12

**CHAPTER THREE**
Mathematics and Classical Reality   22

*PART TWO  -  REALITY AND THE*
*QUANTUM WORLD*

**CHAPTER FOUR**
First Encounter with the Quantum World   51

**CHAPTER FIVE**
Quantity and Reality   65

**CHAPTER SIX**
More about Physical Quantities   76

**CHAPTER SEVEN**
On the Extent of the "Lingua Mathematica"   86

**CHAPTER EIGHT**
Virtual Processes   95

**CHAPTER NINE**
Back to Classical Reality   105

CONTENTS

**CHAPTER TEN**

Decoherence   114

**CHAPTER ELEVEN**

Did You Say "Paradox"?   126

*PART THREE  -  THE CHARACTER OF PHYSICAL LAWS*

**CHAPTER TWELVE**

The Character of Fundamental Laws   141

**CHAPTER THIRTEEN**

The Character of Classical Reality   164

*PART FOUR - PHYSISM*

**CHAPTER FOURTEEN**

The Philosophy of Mathematics   179

**CHAPTER FIFTEEN**

Physism: The Thesis   199

**CHAPTER SIXTEEN**

Physism and the Philosophy of Mathematics   216

**CHAPTER SEVENTEEN**

Physism: A Discussion   231

**CHAPTER EIGHTEEN**

Ontology   243

BIBLIOGRAPHY   253

INDEX   261

# Preface

There never were so many people doing or using mathematics, and never was mathematics so essential in science and technology. Nonetheless, the old question: "What is mathematics?" remains a mystery. The nature of this strange science is still one of the most intriguing problems in the philosophy of knowledge. Some people marvel at mathematical beauty, but that is not really a problem, because we can never explain why something is beautiful. The real questions have to do with the extraordinary fecundity of math, unceasingly generating new concepts, new patterns, and solving or raising new problems on which novel constructs can grow. One does not need to know much to wonder why distant parts of math can join together in perfect harmony to produce an unexpected progeny, or why they suddenly become mirror images, like geometry and commutative algebra. Fecundity and harmony (or rather, consistency) are the main unexplained features of pure mathematics. One should add to that an intimate relation with nature, and particularly physics, whose basic concepts and laws cannot be disentangled from mathematics, particularly in quantum mechanics and general relativity. One could also contrast the axiomatic roots of mathematics, so elegantly pure, with the multitude of its branches. Surely, there must be a simple answer to the question of the nature of mathematics, but which one? Could it be obvious? Could it be that Aristotle's famous saying, that our minds are blind to what shines most luminously, is true once again, and the answer is before our eyes?

There is no lack of tentative answers. A famous old one, still in much favor, goes back to Plato. It says that there exists another kind of reality, an ideal world in which the mathematical ideas are dwelling, a reality that our minds are able to discover. An objection almost as old however, asks, how our brains could get a glimpse of that foreign world! Many mathematicians, philosophers, and biologists, on the contrary, emphasize the human side of math. The possibility of inventing it would just be another example of mankind's abilities, similar if not identical to other adaptive accomplishments of the human species. Brain science certainly makes that opinion credible, but it does not explain why this game, which we are supposed to have invented, possesses an autonomy, a consistency, and a vastness that we did not put into it, why it resists our efforts until we find a key that always seems to come from it and never from us. There are many versions of these two basic attitudes, which insist either on the self-sufficiency of mathematics or its invention by human beings, but none of them explains acceptably why mathematics is fecund, consistent, and in agreement with nature. The situation is confused, and Reuben Hersh has shown nicely in a recent book how each main school in the philosophy of mathematics singles out one or two features of math, ignores others, and can never make clear why another approach explains the aspects it has left aside (Hersh 1997).

Another school of thought often prevailed, before the famous crisis in the foundations of mathematics, at the turn of the nineteenth and twentieth centuries. It appeared historically just after Plato's proposal, when Aristotle attributed the origin of mathematics to an abstraction of the features of nature. Our mind is able to list many features of a real object and turn it into a purified paradigm, with which our reason can play. Most historians of mathematics agree that numbers and geometry had this origin, and the direct extraction of mathematical ideas from an observation of nature was still at work when calculus was invented in the seventeenth century.

Philosophy hesitated after that. Some philosophical worries led Descartes to renew Platonism, and in contrast different concerns led Leibniz to propose formalism (in which math would be a free game of assumption and deduction). Both of them tried to understand what is "a" triangle in a theorem, something encompassing all the possible triangles though not reducible to any that we can draw. Kant's schematism boiled down to saying: Just consider how a triangle is produced and do not think of it as something given from outside. Look how you obtain it by drawing three straight lines with a ruler, and then you abstract the construction process, not the object. Algebra fitted this pattern nicely, when it was applied to calculations, and one was then back again to the abstraction of nature, except that actions were also taken into account. Marx and Engels were more drastic, although they did not know the mathematics of their time. They reduced mathematics to a "superstructure" of matter, a notion that could be understood only with the help of the peculiar dialectic of Marxism.

The link between mathematics and nature weakened with the growing abstraction of mathematical concepts during the second half of the nineteenth century, and it broke—at least among philosophers and mathematicians—when the "crisis," initiated by Cantor and Frege, occurred. The chasm continued during the twentieth century, and I will show in a moment why physics contributed to widening it. After Henri Poincaré, no mathematician and no philosopher of influence proposed associating the roots of physics and mathematics in a common pattern.

The question of this book will be, nevertheless: Can one still conceive of the existence of mathematics as a consequence of the laws of nature, and especially of the laws of physics? When asking that question, one does not assume, of course, that the laws are present in the mind of a mathematician when he or she is working. One does not even suppose that this person often gets inspiration from something belonging to nature. But when one is writing a book on such a topic, the necessary precautions of that sort can be left for the main text. The aim of a preface can only be to hint at a few simple and suggestive ideas, and that's what I will now try to do.

What about consistency, for instance, which is the truth criterion in mathematics? Is there anything analogous in physics? No epistemology book fails to mention the role of experimental verification of a theory in the natural sciences. After Karl Popper, one recognizes a philosopher by his or her penchant for the word "falsification" in this connection, meaning that the theory is sacrificed if experiment condemns it. Popper was much impressed by Eddington's observation of the bending of light rays when they grazed the Sun during a total eclipse. Einstein had predicted this bending as a consequence of the relativistic theory of gravitation and, had it failed to occur, the whole theory would have been of no value. Popper's criterion of scientific truth is well taken, of course, but one often forgets that it is not unique. Nobody would have considered general relativity as a serious theory if it had not been, first of all, a consistent mathematical construct. In the year 1920, when the experiment was performed, the consistency of the theory had not yet been checked in full detail, but one may presume its credibility from the fact that Hilbert himself had worked at it seriously. Anyway, let us stress the fact that a good theory must not only agree with experiments, but must constitute also a good piece of mathematics in order to make sense.

How far does this requirement of consistency extend, in physics and other natural sciences? I would say as far as in mathematics itself. If a physical theory relied on a nonconsistent piece of mathematics, it would obviously be of no value and would be falsified before any experimental check. The strength of consistency as a truth criterion is most conveniently appreciated when the mathematics in a physical theory has already reached an axiomatic form. The consistency requirement, if it were asked only of that theory, would then extend automatically to everything necessary for the construction of the theory, and particularly to the relevant mathematical axioms and all the mathematical consequences of these axioms. This means that if a physical theory belongs formally to a definite domain of mathematics, both the theory and the associated mathematics would be falsified together through any lack of consistency.

The simplest example is the oldest one: counting objects (coins, cattle, and so on) was essentially the ground floor of science. It involved only the recognition of a similarity among different objects together with the abstraction of their difference. It led to the basic operations of arithmetic in a straightforward way, with rather deep consequences. To give just a simple example, one can mention the problem of dividing the cattle of a dead father into equal parts among several sons. It leads immediately to the idea of prime numbers. Fermat's last theorem is then not far away. Quantum mechanics provides a wealth of consistency requirements, because of its deep mathematical background: It relies upon almost everything necessary for the mathematical theory of Hilbert spaces (because of wave functions and spin), distribution theory (because of the delta function), spectral theory (because of measurements), pseudo-differential calculus (for the correspondence between quantum and classical physics), and a lot of group theory. When these pieces of mathematics are decomposed into their basic structures and axioms, as Bourbaki would do it, one finds that everything in the theory of sets, including the axiom of choice, belongs to the consistency domain of quantum theory.

Consistency will be further discussed in this book, but another essential feature of mathematics is its fecundity, which is indeed stupendous, although somewhat less impressive when compared with the fecundity—or creativity—of physical laws. What I mean by the creativity of the laws is as follows. Suppose one writes down the few equations defining the standard model of particle physics, together with the underlying quantum laws. They can be contained in one or two pages. One should of course acknowledge that they are the foundations of physics as we know them presently and they remain open to improvement, but let us look nonetheless at their distant consequences. They say that many particles, and particularly protons and neutrons, are made of quarks. After some hard work, one understands how that implies the existence of nuclear forces, which give rise to the various species of nuclei. The nuclei can bind with electrons—because of electric coupling, included in the standard model—and that implies the existence of atoms, with all their properties. A remarkable property of that kind is the existence of chemical laws, so that a grandchild of the standard model after these many steps is a theory of molecular binding. Then comes a very remarkable effect, decoherence—on which more will be said later—which turns the probabilistic laws of quantum mechanics into classical ones, when many particles are concerned. Causality, or mechanism, can then occur and dominate in a lot of different wonderful ways in nature, live as well as inert.

It would have been easy to point out many other directions along which the basic laws proliferate, but their multiplicity is obvious. I do not mean, of course, that all the steps of creation (an actual creation throughout the history of the universe) are understood in detail, but if one remembers that the human brain is one of the latest outcomes of the process and one of the most far-fetched consequences of the basic laws, the fecundity it discovers when doing mathematics should certainly be no surprise. How a human mind can invent new mathematics when the suggestion does not come directly from nature, how this mathematical product turns out to be consistent with the rest of knowledge, is another question raising fascinating cognitive problems, which will be sketched in this book. The problem of the fecundity of mathematics does not seem in any case a serious difficulty for the present approach.

I mentioned already the remarkable correspondence between various parts of mathematics as something that must be explained. One of the most remarkable examples goes back to the 1950s when Alexander Gelfand and Ian Segal discovered independently an equivalence of geometry with some higher algebraic structures: the so-called commutative $C^*$-algebras (pronounce $C$-star). I leave aside the definitions and restrictions that are necessary for an exact statement of the corresponding theorems, but this is a wonderful case of correspondence, where genuine spaces (a sphere, an ellipsoid, anything) arise from the algebraic rules of abstract objects. The converse result, namely, producing commutative algebras by considering the functions on a space, is much easier. This is not the end of the story, however, because this equivalence of geometry and algebra was already familiar in quantum mechanics, although in a rough way, the corresponding key words being "configuration space" for the geometric object and "complete set of commuting observables" for the algebra. I do not want to dispute whether physicists or mathematicians discovered the fact first—mathematicians in any case made it more cleanly—but my point is to stress that the harmony between apparently distant parts of mathematics looks much more likely when one can compare it directly with an inner harmony of the laws of nature.

I suggest the name "physism" for the philosophical proposal that considers the foundations of mathematics as belonging to the laws of nature. It should be stressed that the root in this neologism is not the science of "physics" as taught in the physics department of a university, but refers to the Greek word *physis* or *phusis*, meaning "nature." The name is patterned on "logicism," a well-known philosophy of mathematics (perhaps using the expression "physicism" would make the analogy closer, but it already has a different meaning). This proposal raises a few new problems in the epistemology of mathematics, which I will try to discuss. An encouraging consequence, anyway, is that it brings the previously well-known philosophies of mathematics much closer together, and it explains reasonably well the partial success of each one of them as well as their common failure.

But surely, there must be a reason why physism is not already among the standard approaches to a philosophy of mathematics. The best guess is probably a historical difficulty between mathematics and reality, which had two successive phases. During the first phase, which started near the end of the nineteenth century, neither Hilbert nor Russell, nor even Brouwer, thought of math as an image of reality. They were either influenced by a vision where pure mathematics was supposed much wider than plain physical reality, or they were convinced of the precedence of logic. When physical reality began to reveal its hidden wealth with the discovery of relativity and quantum mechanics, the situation could have been drastically changed, but physics unfortunately appeared then a shaky basis for constructing a unique palace of knowledge.

This point is extremely important for the history of ideas, at least in my opinion, and I wish to make it clearer. One may start conveniently from the idea of reality, which was among Kant's categories of reason and was endowed by him with several essential characters: uniqueness, rationality, continuity, a spatial framework, and a causal time evolution of its content, with a clean separation of the various real objects. Continuity was the first property that was questioned, with the discovery of quanta in 1900. The space-time description was, of course, much perturbed by relativity, after 1905. The discovery of the laws of quantum mechanics, around 1925, had much more drastic consequences. Causality was negated, because of the randomness of quantum events. The discovery of Heisenberg's uncertainty relations—which result from the quantum principles—disposed of the space-time representation of reality, since one could not *think* of a particle as having a definite motion in space. Schrödinger noticed also that, sometimes, it is impossible to disentangle different objects from each other, even when they are in different places. Von Neumann and Schrödinger encountered dire difficulties when they tried to reconcile the uniqueness of empirical reality with the foundations of quantum theory, under the conditions of a quantum measurement. Finally, the rationality of reality also became questionable in view of the famous wave-particle duality and Bohr's introduction of "complementary" concepts in a quantum description. Complementarity implied, for instance, an incompatibility between two different though simple and reasonable concepts, such as describing light as a wave or as made of photons, two incompatible descriptions though both necessary under different circumstances. Not much of reality remained after such a slaughter, and one cannot be surprised that philosophers and mathematicians alike did not consider favorably the perspective of looking at physical reality for an inspiration, when they were still struggling with the meaning of mathematics.

Significant progress has been accomplished, however, in the understanding of quantum mechanics during the last few decades. The most important discovery is the existence of the decoherence effect, through which the quantum laws are transmuted into classical ones. That does not mean that all the problems are solved, but physics has clearly returned to objectivity. The notion of reality is not yet clean, and one should certainly not envision building anything firm on it right now, but that is not the proposal of physism.

The laws of nature provide us with something more secure than reality, even if that looks like a paradox. The laws governing space-time and quantum physics have withstood a tremendous increase in scope and in precision over nearly a century. One must acknowledge, of course, that our knowledge of a law of nature can evolve; it may become deeper and change in conceptual emphases, but nobody is seriously considering that the laws we know could be wrong. They are certainly incomplete, but wrong? That's impossible! This is why this book will strongly rely on the character of physical laws, as they are known by now. They will be analyzed and, as a general rule, analysis—or critique in the philosophical sense of the word—will prevail over speculation in this work. This is why the questions concerning physical reality will mostly be left out, with the fortunate circumstance that they do not seem necessary for clarifying the problem of mathematics. The ontological question (regarding particularly mathematical Platonism) will mostly be omitted, not because of a lack of interest but because the answer is inaccessible to an analytic approach.

Is the present proposal original? Perhaps, but who cares? The important point is whether it holds some amount of truth. I rather consider it myself as something in the spirit of the time. The recent renewal of combined research in physics and mathematics is an indication of the trend. The growing importance of mathematical consistency in fundamental physics is another, as seen, for instance, in the historical development of the standard model of quarks and leptons. The ambitious programs of noncommutative geometry and string theory obviously rely on the consistency criterion of truth in physics and its presumed fecundity. I did not try to press this similarity, however, because I wanted to insist on an analytic approach, independently of any beautiful but still speculative enterprise. If one of these promising approaches succeeds, one may expect that its philosophical background will be much investigated, but the time is not yet ripe.

Finally, I must become personal and say that I entered the field of philosophy of mathematics, where physicists are not often invited, because of a long-standing interest in the meaning of physical laws. I was told I could thus risk my reputation, whatever it is, but passion is sometimes stronger than concern for one's reputation. A difficulty is that I cannot claim scholarship in the philosophy of mathematics, and that lack will certainly be exposed in this book. I shall not argue a life-long interest in these matters in my defense: that cannot turn an amateur into an expert. Neither will I take as an excuse that most people, in this time of specialties, encounter similar drawbacks. I only want to explain why I am not in a position to write a scholarly book and this one must be an unpretending essay, which I tried only to make as clear as I could.

A few years ago, this book was supposed to be a translation of a French one I had written on the same topic. I am inclined by now to compare the way it was cooked up with a Burgundy recipe, when a pheasant is stewed inside an envelope of veal flesh. The veal is thrown away at the end of the process. My French book was cooked in an envelope of ontology, but, contrary to gastronomical rules, that was not a good recipe. Ontology did not have the right taste and it made the meal too fat. In the end, the present book is completely new, and I hope it is a better one, if only because it is now able to withstand Occam's razor with no unsavory leftovers.

## Acknowledgments

I benefited from many useful remarks and criticisms when writing the successive versions of this book. The most pungent ones suggested that nobody (and myself particularly) could know enough of natural sciences, mathematics, and philosophy to avoid making a fool of oneself with such an enterprise. I could not disagree, but I held that if the idea could get out, knowledgeable people would perhaps examine it separately under its different aspects. This is why I thank here some of the people who encouraged me and made me dare to publish this book in its present form: Robert Dautray, Jean-Pierre Kahane, Jean Petitot, Simon Saunders, and Andrew Wayne. Georges Jobert suggested "physism" as the most convenient expression for the present approach, much more convenient that "physicalism" or "physicism" (introduced in French by Saint-Simon in 1808), which usually means a reduction of the laws of nature to those of physics and would be misleading in the present context. I also thank Joe Wisnovsky who gave to this work the prestigious hospitality of the Princeton University Press. Patricia Flad helped me with bibliographical research. Jennifer Slater did a magnificent job in polishing my English.

# PART ONE

## PRELIMINARIES

# Beginnings

### THE EARLY DAYS

**M**any findings in archaeology bear witness to some math in the mind of our ancestors. There are many scholarly books on that matter, but we may be content with a few examples. A bone rod, which was discovered in 1937 in Moravia, shows 55 notches in groups of five and is about 30,000 years old. Paintings on cave walls and many engraved objects show various forms of geometric design, more and more sophisticated as one approaches the beginning of the Neolithic period (12,000 years ago, when agriculture began).

Mesopotamian, Egyptian, Indian, Chinese, and Old American civilizations already knew as much math as is taught in the first grades of our high schools. Basic arithmetic, including the four operations, is carefully described in Babylonian tablets and Egyptian papyri. The many exercises accompanying the descriptions show that mathematics was then an empirical science: the examples are very practical, showing, for instance, how one can divide a herd into so many equal parts (quite useful in a case of inheritance or when sharing some plunder). Prime numbers then begin to appear. Mesopotamian, Egyptian, and Chinese scribes knew much geometry. Figures 1.1–1.3 give examples of their knowledge. Figure 1.1 shows how the Egyptians computed the area of a

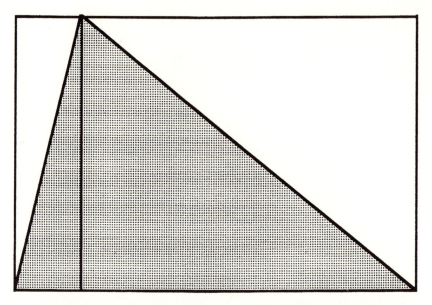

Figure 1.1. How the early Egyptians computed a triangle area. This drawing shows that the triangle area is half the area of the rectangle built on its base and height.

triangle. Herodotus, the Greek historian, explains that Pharaoh's administration taxed land on the basis of acreage, and one sees how practical this kind of recipe could be. Babylonians knew the so-called Pythagorean theorem (involving the sides of a rectangular triangle), probably on the ground of some drawing like figure 1.2 (figure 1.3, which is similar, is from a later Chinese source). One could also mention good approximations of $\pi$, the volume of a pyramid, the solution of second-degree algebraic equations, and a few other items, but they would not add much to the basic statement: much mathematics was known early and always for practical reasons with empirical means.

Figure 1.2. A drawing yielding the Pythagorean theorem. By moving triangles, one concludes that the shaded square area in (a) is the sum of the two square shaded areas in (b).

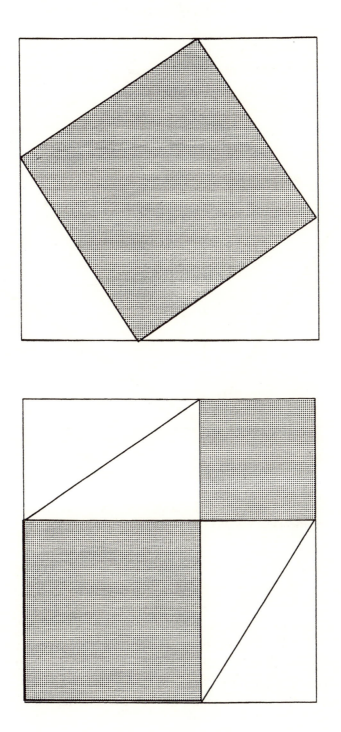

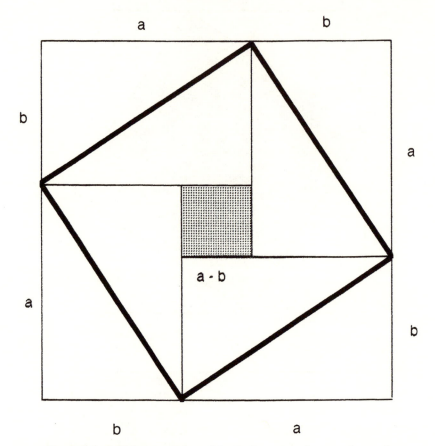

Figure 1.3. A Chinese version of the Pythagorean theorem (around 200 B.C.). If $a$ and $b$ denote the sides of a right triangle, the boundary square has a side $(a + b)$ and the smallest shaded one a side $(a - b)$. The side $c$ of the blank triangle is obviously such that $c^2 = (a - b)^2 + 4(ab/2) = a^2 + b^2$.

## THE BEGINNINGS OF GREEK MATHEMATICS

Babylonian and Egyptian texts show that the early mathematicians were working on bookkeeping, agronomy, architecture, or astronomy, but one does not know what dreams they entertained around their art. Greek mathematics begins, on the contrary, with the flamboyant figure of Pythagoras (c. 580–500 B.C.).

Had he lived one century earlier he would have come to us as a legendary personage, like Orpheus or Theseus. Myth had already begun haloing him, since he was said to have a thigh made of gold. He founded a quasireligious sect that still existed in Plato's time, 150 years later. Little is known of his doctrine, but he certainly held that "numbers govern everything," whatever that means; he was eagerly interested in mathematics and he is said to have sacrificed an ox when he discovered (or perhaps proved by new means) the Pythagorean theorem on rectangular triangles.

The earliest mathematicians were certainly Pythagoreans and an interesting speculation is mentioned in Bourbaki's *Elements of a History of Mathematics*, which would make math begin like a novel (von Fritz 1945). It may be entertaining to assume this story is true, as follows: One of the main symbols of the Pythagoreans was the pentagram, the stellar regular pentagon that was always treasured by all sorts of mystical groups. They knew how to construct it inside a circle with a compass and a ruler, and the construction shows easily that the ratio of the stellar pentagon side to the radius of the circle is the "golden ratio" $a = (1/2)(\sqrt{5} + 1)$. This ratio played a great role in Greek aesthetics, including painting, sculpture, and architecture, and it was certainly one of the numbers governing the world, according to the Pythagorean doctrine. This doctrine held, however, that the numbers worth considering are integers or made of integers. The golden ratio should therefore have been a quotient $a = p/q$ of two integers $p$ and $q$. One could divide the two of them by a common divisor until neither was left. And then comes the drama! The obvious relation $a = 1/(a + 1)$ implies that $p(p + q) = q^2$. But that means that $p$ and $q$ must have a common divisor, which is impossible. Something was then wrong in Pythagoras' mystical assertion.

Of course, this story is invented or at best speculative, but one knows for sure that some Pythagoreans investigated the diagonal of a square and proved without any possibility of escape that $\sqrt{2}$ cannot be rational (i.e., a quotient of integers).

That was a hard blow for Pythagoras' doctrine. It occurred probably early in the fifth century B.C., but one does not know who found it. Did one man discover it, or maybe a woman, since the sect accepted women? Or was it the outcome of long investigations by a group of people? We don't know. We know that a man, Hippasus of Metapontium, was accursed for letting the secret of the result leak out, but that does not mean that he discovered it.

On the other hand, one cannot underestimate the importance of the event. It was certainly the first true theorem, the first one at least that could not be made obvious by means of a clever drawing. It implied to Greek eyes that the mind is able to reach a hidden truth by itself, using only the power of its own thought. It revealed the power of logic, which was still in the turmoil of initial searches. It increased the Greek confidence in the supremacy of ideas and, incidentally, it also led for a long time to disparagement of the empirical approach to science. Last but not least, it showed the value of rigor, which led the way toward Euclid's axiomatic construction of mathematics.

## PLATO AND THE PHILOSOPHY OF MATHEMATICS

There is no indication of a specific philosophy of mathematics after the failure of Pythagoras' attempt. The main questions were concerned with general philosophy: what is Being, and should there be non-Being, is there infinity, what are Good and Beauty, what is reason (although people preferred speaking of the nature of ideas), what is life, and many other such issues. Mathematics entered that game with Plato in a rather roundabout way, more as an example than for its own sake, but it was the beginning of a long story.

Plato (c. 428–348 B.C.) knew well the mathematics and the mathematicians of his time, including particularly Eudox, who was his student and friend. One of his main interests was, however, the meaning of ideas. He often took mathematics as a paradigm and one of his early dialogues, the *Meno*, gives a good indication of his early views on that matter. At some point in the dialogue, Socrates asks a boy various mathematical questions such as: How much larger is the area of a square when the side is multiplied by two? How big is the square built on the diagonal? The boy is supposed to know nothing, since he was born a slave. . . . Socrates shows himself, however, a very kind inquirer, he gives many clues, he suggests which lines can be drawn to get a hint, and every student would certainly get an A grade with such an examiner. The boy answers correctly, of course: he is led so gently, and a modern reader would simply conclude that he does not lack common sense. That is not Plato's conclusion, however. The answers prove that the boy knew them before he was asked the questions, and the query only helped him to remember them!

This example was typical of Socrates' method, which Plato learned in his youth. The Pythagorean School influenced him later, and mathematics then became a more important key in his philosophy. His main assumption was the existence of two different worlds. There is the world we see with our eyes and in which our body is immersed, a world that can be considered as more or less trivial according to Plato. There is also another world in which perception is replaced by understanding, and mathematics originates from it. Our senses cannot reach that world, which is supposed to be more real than the one we live in, and its inhabitants are immaterial. They are the Forms, or Ideas. They belong to the sphere of divinity, higher than the gods themselves, and they share a common harmony including everything that is "Good."

Plato considered, for instance, that the world of Ideas contains a Form "circle" embodying all the possible circles in the world below, and similarly a Form "triangle." He was much impressed by the fact that mathematical properties can be discovered though nothing hints at them in the definitions, the fact, for instance, that the median lines of a triangle meet at a common point at one-third of their length. That property was already there before any worldly triangle had been drawn for the first time. This feeling of a preexistence, of something more real than reality, has always impressed many mathematicians, and one may presume that Eudox, who was one of the greatest mathematicians of all times, confirmed Plato on that point. There are still many believers or would-be believers in this form of Platonism among modern mathematicians, who feel something like the existence of another world where mathematical truth rests.

Plato was aware of an obvious objection to his proposal: How can we, we people made of flesh, who live in this world down here, how can we get in touch with the ideal world where mathematical truth dwells? His answer was that our soul inhabited that world before we were born, and we have memories of it. We may of course forget this answer, but the question itself will remain interesting.

## ARISTOTLE AND ABSTRACTION

Aristotle (c. 385–322 B.C.) brought mathematics back down to earth. He considered the mathematical objects, numbers, circles, triangles, and so on, as so many abstractions of real objects, either natural or manmade. Although every line we draw with a stiletto on a waxed plate has a finite breadth and irregularities, our mind can make an abstraction of them, forget them, and consider them as irrelevant. Irrelevance is the motto when somebody worries about giving the same name to two obviously different triangles: everything making them different is inessential. Aristotle in that sense considered the mathematical objects as very close to natural objects, or at least as patterns that are found in reality.

Plato's problem, "Why are there mathematical properties that are not contained in the definitions?" received a new answer: Logic can create new truth, and this kind of property gives a perfect example of its ability. One should not forget

10

that the discovery of logic was still recent, and Aristotle was one of its major investigators. Some of his concepts are worth mentioning and, rather than choosing them in his *Logic* or his *Metaphysics* to which I intend to return, I will pick them up in his *Physics*. He says in that book that we cannot really understand something without knowing its first principles. He enters then into various predicaments about Being and non-Being, about motion as a transition from being there to not being there. He states as a principle that every motion must proceed from a permanently active cause (a principle that, by the way, impeded physics for a millennium and a half), so that a moving object is moved by another, which is also moved, and so on, until one must arrive finally at a primary mover, who pushes the sphere on which the stars are nailed in a perfect motion that is necessarily rotation. Aristotle's book is a work of beauty and also an ascetic song of love for nature, *physis*, since love and hate are among its other basic principles: a stone falls because of its love for the earth and smoke rises up for love of the sky.

Philosophers enjoy that book for the tension in its argument and they do not worry that most of its conclusions have turned out to be wrong. Physicists would rather say that it is not a physics book in spite of its title. I wished to mention it however, because of its relation to the main topic of the present book and particularly in view of two significant statements by Aristotle, namely, (i) mathematics relies primarily on an abstraction of reality; (ii) physical reality can be understood only by getting at its first principles. These statements could look like our thesis of physism in a nutshell, except that mathematics, the principles of physics, and even the meaning of reality have much changed in the meantime.

## CHAPTER TWO

# Brain and Reality

### ABSTRACTION AND THE BRAIN

The essence of mathematics appears to be a study of the *relations* between some objects, which are (voluntarily) known and described through only *a few* of their properties" (Bourbaki 1960). This strong statement implies that abstraction is the keyword in math, a dominance that is made still clearer when Bourbaki adds that the "few properties" to be retained in the study of mathematical relations are "the axioms at the basis of the theory." Abstraction is therefore supposed to dig deeper and deeper and to strip mathematical objects of more and more of their properties, until some ultimate axioms are reached.

The purpose of the present chapter will therefore be to study abstraction. It will not be considered abstractly however, but plainly, as something our brain does more or less casually. Among the many trends in the philosophy of mathematics, we begin therefore with one of the latest: the cognitive approach, which reminds us essentially that math is made by the mathematician's brain.[1] "Mathematics as we know it is . . . a product of the human mind. . . . It comes from us! We create

[1] See, for instance, Changeux's contribution in Changeux and Connes (1995).

it, but it is not arbitrary [because] it uses the basic conceptual mechanisms of the embodied human mind as it evolved in the real world. Mathematics is a product of the neural capacities of our brains, the nature of our bodies, our evolution, our environment, and our long social and cultural history."[2] These statements would have looked trivial and useless a few decades ago, but the present advances in brain science begin to make them helpful: To know, for instance, that abstracting is one of the most commonplace performances of the brain is certainly something new and significant. Psychologists would have considered it earlier as a highly evolved operation, and that makes things very different.

The present chapter will draw much, therefore, from discoveries in neural science. It is strongly influenced by the cognitive approach to mathematics, but it will also draw our attention to the obvious fact that the brain is something real, perceiving real things, so that reality is prior to cognition. Well, doesn't all that sound trivial enough? It's time to show that there is more in it than meets the eye.

### THE QUESTION OF REASON

The ancestors of the cognition scientists were John Locke and David Hume, and they were not much interested in mathematics. Their empiricism, however, which wants to explain the origin of ideas and language as well as the meaning of reason, will often provide us with a useful background. Ideas, for instance, according to Hume, "are copies of our [most lively perceptions.] . . . Every idea is copied from some preceding impression or sentiment." Hume also noticed that "though our thought seems to possess [an] unbounded liberty, we . . . find, upon a nearer examination, that it is really confined within

---

[2] Lakoff and Nuñez 2000, 9. (This quotation is borrowed from Henderson 2002.)

very narrow limits and that all this creative power of the mind amounts to no more than the faculty of compounding, transposing, augmenting, or diminishing the materials afforded us by the senses and experience." Concerning language, one may quote Locke: "Words become general by being made the signs of general ideas; and ideas become general by separating from them the circumstances of time and place, and any other idea that may determine them to this or that particular existence. By this way of abstraction they are made capable of representing more individuals than one."

Roughly speaking, empiricism considers that reason results from an adaptation of human beings to the universal regularities in nature. Language is also a by-product of these regularities, a wonderfully clever way of expressing the information from them. Reason and language proceed through an abstraction of differences among individual cases and a selection of their common patterns. Some other aspects of early empiricism have become partly obsolete—its reduction of physical laws to a constant habit, for instance. But we had better leave out general philosophy at this point.

The cognitive sciences have much refined empiricism. They take advantage of many contributions by philosophers, linguists, mathematicians, computer scientists, physicists, and of course, prominently, biologists and physicians. An essential early improvement was to take evolution into account, so that millions of years since the origin of humankind became available for explaining the development of language and reason, rather than the short span of a human life. Evolution grants, moreover, hundreds of millions of years, since every animal species, including bees, whales, and chimpanzees, has a form of language using sounds, gestures, or smell (this line of research is also presently very active). Although one cannot speak properly of reason, every animal is able to gather information from the regularities of its surroundings and utilize it (the name "Information Gathering and Utilizing System," IGUS for short, has been coined for this by Murray Gell-Mann). The scope of the

14

idea is even wider since it extends to plants and bacteria, thus allowing billions of years for the slow rise of reason. Our search for the meaning of mathematics is of course narrower, but it can ignore neither brain science nor the framework of nature, from which every kind of information is gathered.

## Perception

The brain gains information from reality through perception. Brain science by now provides an impressive amount of data on perception and its mechanisms, but it will be enough to mention a few significant results concerning vision, which is the most convenient example. One knows from physics how an object emits light or reflects it, and the laws of optics can also explain how light produces an image on the retina after crossing the optical part of the eye. Then perception really begins. The retina involves 140 million photoreceptors, which are surrounded by an interconnected net of neurons. These neurons are essentially an outside part of the brain; they send their visual information to the cortex through the optical nerve, which consists of a million or so long fibers (or axons) belonging to a family of transmitting neurons. Each axon carries electrical pulses, and the collection of these signals provides a message, which is sent to the cortex and provides it with all the information that the brain will have about the image. One might say that this message is coded, since the initial optical data have been transformed into electric data, somewhat analogous to the electromagnetic signals of a television network.

It was believed for a long time that the retinal neurons transmit the image directly to the brain and preserve its shape and color exactly, but they actually perform a much more elaborate work of filtering and coding. As a matter of fact, the information inside the retina is too large to be transmitted usefully: We saw how huge the number of photoreceptors is and

15

there are about as many neurons, each neuron emitting and receiving pulses. This accumulation of little signals would produce a mess if it arrived directly at the brain and, furthermore, it could not be carried through the few fibers (a mere million) in the optical nerve. Some order must be introduced and it comes from synchronism: It has been found that many neurons can cooperate when they adjust their pulses to an exact simultaneity, while the random pulses originating from noncooperating neurons have no significant effect. The information reaching the brain is only the outcome of large bunches of synchronous signals originating from cooperating families of neurons.

The synchronous action of neurons has an immediate consequence on our representation of reality. Without knowing how the brain constructs this representation, one may be sure that it must be governed by time, ordered in time, and practically instantaneous, since the information generating it has this character. This biological priority is obviously reminiscent of Kant's vision of time as a "pure form of intuition": something intrinsic to reason. He went too far, however, when he excluded time from reality and regarded it "as a condition of the possibility of phenomena, not as a determination produced by them." It may be important to recognize, anyway, that we possess an embodied *intuition* of reality, which does not necessarily reflect the true features of reality and remains a result of the work of the brain.

The collective mechanism of neurons also has a consequence on our intuition of space. It has been observed that various groups of neurons working together are associated with small circular regions of the retina. As a consequence, the retinal image is spread over these regions when the information arrives at the brain. Perception therefore enforces a *continuous representation* of reality, which is not intrinsic to the retina itself, since each photoreceptor can detect light photon by photon. This detailed information is thrown out, however, when many neurons adjust together to emit synchronous signals. It may be added that, when we recall the memory of an image, the zones at work in the cortex are the same as when the image is actually seen. Our imagination is therefore biologically constrained to continuity, and this simple observation will enlighten a few significant points in the history of mathematics when we come to that.

The signal incoming from the retina is analyzed in more than thirty regions of the brain, each one having a definite localization and a specific function. Some of them recognize straight lines, and different zones react to horizontal or to vertical lines. Other regions are specialized in the analysis of color, or in detecting motion, or various other aspects of images. The different regions also exchange their information during the process, and there exists a complex of regions where the recognition of objects takes place. The theory in best agreement with experimental findings assumes that this pattern recognition involves a comparison by the brain of what is actually seen with what has been previously memorized. Memory is not perfect, however, and the brain remembers only, for instance, some features of a tree it has seen and not every detail. If one puts together this imperfection of memory with the decomposition of an image into various component by the cortex, one may fairly presume that abstraction is one of the most fundamental and ordinary process of the brain. This is certainly interesting when one turns to mathematics, and also a good point in favor of Aristotle against Plato.

What about consciousness? It is poorly understood but some experiments show a few interesting features in it. One may project, for instance, two different images that are seen by the two eyes of a cat: the left eye sees an unmoving bee and the right one a moving butterfly. A mirror standing right on the nose of the animal makes its brain see two superposed images, because of binocular vision. Electrodes in the cat's brain indicate what he is conscious of seeing (or at least where his attention is caught): if the left eye is seeing consciously for instance, the neuronal signals emitted by the cortex regions that are associated with that eye become synchronized, whereas the signals originating from the right eye do not. In these conditions, the brain never shows a simultaneous awareness of the two eyes, and only one of them is responsive. This is to be contrasted with the simultaneity of other unconscious perceptions, indicating for instance that the bee is perceived though not seen when the cat's attention is concentrated on the butterfly. This observation is very important, since it indicates that consciousness—or awakening, or whatever it is—implements uniqueness in the brain's representation of reality.

A last datum from neurology is also relevant for our topic. It is concerned with pattern recognition and, once again, I will give only one example. Figure 2.1 shows a drawing that

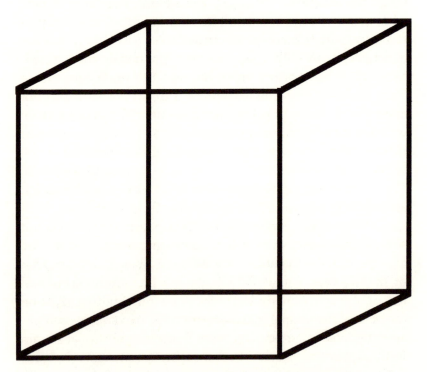

Figure 2.1. An optical illusion. We see the drawing as representing a cube, but cannot tell which face is forward or backward.

our brain recognizes spontaneously as the image of a cube. Because of a lack of perspective, we cannot tell, however, which face of the cube is forward and which is backward. A remarkable physiological fact is that our consciousness always decides which is which, but the decision can switch from one choice to the other: one face of the cube seems to stand in front at some time and a moment later, suddenly, the opposite face pops up in its place. Many similar experiments have been made under various conditions on humans and monkeys, including some with the help of magnetic resonance imagery, and they agree on a common conclusion: Pattern recognition is a universal ability in higher species, and the brain systematically selects a unique pattern at a given instant.

Our brain always separates a scene into definite objects and rejects ambiguities, with no intervention of the will. Sometimes there is interplay of ambiguities and spontaneous decisions, as when we hesitate about the exact nature of an object far away or badly illuminated. The phenomenon fascinated some Italian painters during the Renaissance period and they produced clever drawings and paintings compelling the spectator's consciousness to see different patterns alternately: a man's face or a garden, a bunch of leaves in a tree or a cat waiting, or many other illusions. Salvador Dali, in recent times, pushed the game very far in some of his paintings. There are even cases when the ambiguities of a geometric drawing are so compelling that our brain gets the impression that something in it is moving. The Hungarian painter Victor Vasarely has realized startling paintings on this principle.

## The Categories of Reality and Classical Science

Brain science shows that our representation of reality is strongly constrained by our anatomy and the connections of our cells. If we avail ourselves of an analogy with computers, we might say that the hardware of our brain involves 100 milliard neurons, each neuron having typically ten thousand connections with others. This hardware constrains the software of the brain and its conscious output. Our intrinsic or "biological" representation of reality is thus endowed with several important properties: uniqueness (reality is unique at every moment); localization in time with definite present and past; location also in ordinary ("Euclidean") space (as shown by experimental studies of spatial representation that were not mentioned here). This intuitive representation of reality is continuous and separated into individual objects (never a bee and a butterfly in the same place!). Other investigations, involving actions and not only perception, by humans or higher animals, show an expectative behavior in them, which amounts to an imprint of causality in the brain. One might add also a distinction of reality from the effect of imagination or a virtual appearance, which is manifested in the different reactions of glands and muscles.

Uniqueness, location in ordinary space and time, continuity, separation of phenomena, causality, and a sense of reality: a philosopher would immediately recognize in this list several "categories of reason," as Kant spelled them out. They are constraints of our intuition and not necessarily of reality, as we shall have occasion to discuss. They are not constraints on the abilities of reason, and mathematics, for instance, makes fruitful use of noncontinuous objects. Relativistic physics considers an indivisible space-time, rather than a distinct pair consisting of ordinary time and commonplace Euclidean space. Quantum mechanics, when dealing with atoms and particles, gives up practically every intuitive character of reality. We may thus anticipate that the chasm between our inborn intuitive representation of reality and the concepts of modern science is important for understanding the high abstraction of modern physics and mathematics. It will be convenient in this perspective to introduce a definition of "classical science" that will be used systematically in this book:

*Definition.* A science is called "classical" when its concepts agree with the characteristics of the intuitive representation of reality, namely, uniqueness, location in ordinary space and time, continuity, separation of phenomena, causality, and a clear-cut distinction between the real and the virtual.

## CHAPTER THREE

# Mathematics and Classical Reality

The subject of this book being the relation between mathematics and reality, the present chapter will be devoted to a few landmarks in the history of this relation. The questions to be considered will be in point of fact: What has been the relation between mathematics and the *classical* conception of reality, and how has it evolved? It has evolved much, actually, and that will lead us to recall a few landmarks in the history of mathematics and also a few points about the evolution of physics (I believe, as a matter of fact, that a remarkable discordance between the courses of these two sciences during the last hundred years or so has much to do with the present confusion in the philosophy of mathematics).

### The Elements of Logic

We begin in the third century B.C. with logic, which is inseparable from mathematics. Experts apparently agree that logic began as an abstraction from ordinary language (see, for instance, Kneale and Kneale 1978). Efficient patterns of thought were recognized inside language, so compelling that they seemed able to reach every possible truth. This is at least how logic arose in Greece and how it linked with mathematics in the form we know it. Logic's name means the science of *logos*, which itself initially meant language but was soon endowed with grand ontological connotations.

We saw in the previous chapter that abstraction is a standard process of the human brain, and it was easy for the early thinkers to abstract some patterns of argument from the structures of language. Rhetoric, which is its primary version, had come earlier, and it had been taught for a century or so—for the sake of winning an audience's favor—when logic came out. Aristotle's idea of considering a proposition as an undivided object and denoting it by a symbol (a letter) was an essential discovery, as well as introducing variables in a proposition such as "$x$ is a man."

Negation, which associates a proposition with its opposite (such as "$x$ is not a man" in our example) warrants a brief comment. It was first associated in Greek philosophy with the opposition of being and nonbeing, as some lessons of Aristotle remind us. In his discussion of motion, for instance, an object is presently at some place and therefore not in another place: the one it occupied a short while ago. Motion is thus defined as a transition from being (there) to nonbeing, or the other way round. This approach looks strange at first sight, but it becomes perfectly sound when considered from a cognitive standpoint. An important ability of the mind is indeed to compare a situation that is perceived with another that has been kept in memory, and many experiments have studied this process. The rejection of a proposition as incompatible with another is therefore embodied in our brain, and one of its earliest occurrences during evolution was probably the difference between the prey being there and the memory that it was not there a moment before.

Then there is the principle of noncontradiction, which is acknowledged as Aristotle's most important contribution to logic. It says that two propositions, $a$ and non-$a$, cannot be true simultaneously. It seems also to be imprinted in our mind, but consciousness is essential to it. We saw in the previous chapter the case of a cat that looks at a bee with the left eye and at a butterfly with the right eye. The motion of his eyes and the synchronization of his neurons indicate that, biologically, he can direct his attention only to a unique insect at a given instant. Similar experiments on human beings show that their consciousness requires the same uniqueness of purpose. Our conscious memory has the same kind of constraint, and one may thus assume that the principle of noncontradiction states a property of our brain. One may even go one step further. Some ambiguous situations, such as a cat being forced to see a bee and a butterfly in the same place or a person contemplating a tricky painting by Salvador Dali, are only a consequence of some artifacts: the tricks of an experimenter or the art of a playful painter. This kind of ambiguity never occurs in macroscopic physics, which is the objective origin of our representation of reality, and the fact that empirical reality is unique is therefore the true foundation of the principle of noncontradiction.[1]

Aristotle's elaboration of logic rested on syllogisms, which relied essentially on the existence of categories ($x$ is a man: it belongs to the category of men) or of properties ($x$ is red: it has the property of exhibiting a red color). Categories and properties are obviously related to nature's regularities and, ultimately, they result from the existence of natural laws. Our first conclusion is then pretty evident: the first elements of logic are direct consequences of the laws of nature through the channel of our perception and because of the constitution of our brain, which developed as an information gathering and utilizing system under the government of the laws.

## GREEK MATHEMATICS AND LOGIC

Greek mathematics relied on much fewer concepts than did philosophy. Its universe of discourse—as logicians would say—was well defined and it generated a specific form of logic,

---

[1] It will be seen later why this statement is far from trivial.

somewhat different from Aristotle's system of syllogisms. Euclid's *Elements* bears witness to this logical autonomy of mathematics, but it is only a matter of form. The remarks we made about the inborn origin of Aristotle's concepts obviously remain valid in the case of Greek mathematical logic.

The foundation of Euclid's mathematics on definitions, axioms, and postulates was probably the result of a systematic exertion by mathematicians, who were often also philosophers and teachers. As philosophers, most of them had an inclination for Platonism or a weak form of it, in which mathematics reflected some kind of divine thought. As teachers, they felt secure behind a sure line of defense in front of their paying students, who were fond of asking incisive questions. After two centuries or so, the outcome was the masterpiece of Euclid's books.

The three species of basic assumptions are clearly inspired by reality and our mental representation of it, which relies on the brain's capacity for abstraction. The main definitions of geometry, such as points, straight lines, planes, and circles, are abstracted from nature and art. Conical curves, which are defined as intersections of a cone of revolution with a plane, are tangible. Real numbers were trickier and they were finally defined through an analogy with the points on a straight line. Algebra, which appeared only in the second century A.C., was clearly an abstraction of arithmetical procedures. It was a kind of arithmetic in which the numbers occurring in an operation were abstracted or left free for choice.

Axioms were supposed to be some obvious propositions, such as, for instance, $a + b = b + a$. Evidence, or obviousness, is most often an inability or a reluctance of our minds to think otherwise: when it is not a result of education or belief, it is very close to biological intuition. Postulates were propositions that were found necessary for building mathematics, but not quite obvious. The most famous examples are the postulates regarding parallel lines, and the corresponding misgivings clearly suggest second thoughts about reality itself: how could one reconcile the infinity of straight lines with the concept of a closed universe? Infinity (*apeiron*) had been introduced before Socrates' time by Anaximander, and Zeno of Elea pointed out its apparent paradoxes. His story of light-footed Achilles running after a tortoise is famous: Achilles always needs some time for crossing half the distance separating him from the tortoise at every moment, which means that the race can be split into an infinite number of time intervals. Zeno wrongly associated this infinite sum with an infinite time, whereas the geometric series $1/2 + 1/4 + 1/8 + \cdots$ for the time to cross half, then half the half, and so on, of the distance to the tortoise converges to 1. Infinity was considered mysterious and a matter rather for philosophy than mathematics. Euclid invoked it only in the "method of exhaustion," defining the number $\pi$ by means of regular polygons tending toward a circle. Archimedes was bolder and performed real integrations involving a sum of infinitesimal quantities, but his example remained practically unique for a long time.

## The Golden Age of Classical Mathematics

If a "paradise" were ever intended for the enjoyment of mathematicians (and physicists), it was certainly inhabited during the seventeenth and eighteenth centuries, the time of the Enlightenment, when everything in science was finding harmony: geometry with algebra, algebra with calculus, and calculus with dynamics. The considerations of knowledgeable historians of mathematics (e.g., Bourbaki 1960; Dieudonné 1978) can be summarized by defining this period as *physicalist* and *classical* (when using the word "physicalist" in this circumstance, I do not mean only that physical reality and the problems of physics were the main source of inspiration for mathematics, but also a general attitude toward foundations that will become physism in this book). Both algebra and analysis yielded many remarkable results but they could not receive a safe axiomatic basis, because of the conceptual difficulties arising from complex numbers and infinitesimals. The search for consistency was replaced in most cases by cross-checking the results using different methods and entertaining an overall confidence in the validity of mathematics, due to its deep agreement with physics.

The idea that mathematics derives its consistency from an adequacy with physical reality was never promoted, as far as I know, to the status of a "philosophy of mathematics," but it runs all along the eighteenth century and I propose to call it "physism." It is clearly expressed in two quotations from Fourier (at the very beginning of the nineteenth century): "The thorough study of nature is the most fruitful source of mathematics" and "Mathematical analysis, through its generality and power, appears as a faculty of the human reason, whose finality is to make up for the brevity of our life and the imperfection of our senses."

To say that mathematics was then classical is closely linked with its relation with intuition, which is another aspect of physism. We saw the natural foundations of intuition in the previous chapter and the image of reality entering in mathematical thinking was of course classical, in the sense we gave to that word. Having thus defined the main trend of this long and rich period by physism and classicism, I will not dwell on the proposals that were made at that time in the philosophy of mathematics: Descartes' Platonism, Leibniz's formalism and Kant's schemata, because we will have occasions to return to these topics later.

There is certainly a touch of physism in two of the main discoveries of that period: analytic geometry and calculus. Analytic geometry relied on a realistic interpretation of Euclidean space together with an assimilation of real numbers with the points on a line. It gave a geometric, i.e. a realistic interpretation of the notion of function, "the great creation of the seventeenth century." Functions were mostly represented by mechanically generated or hand-drawn curves, together with a few newcomers including the logarithmic and exponential functions. They were routinely conceived as continuous and, when derivatives were invented, their existence was always taken for granted.

One may mention as an example the case of the cycloid, which is the trajectory of a mark on a tire when the tire is a perfect circle rolling along a straight line and the mark is a point on the circumference. It was the pet guinea pig of some of the brightest minds in history and, clearly, there was a strong correspondence with reality. The cycloid perfectly fitted analytical geometry, according to which a curve is referred to two coordinate axes and is defined through either a parametric expression of the coordinates or an equation linking them. Much ingenuity was used to construct a tangent to the cycloid, then the osculating circle (which is the circle touching the curve most closely at some point), and also the area of an arch of a cycloid. So then, little by little, the main constituents of calculus were discovered, until Newton and Leibniz gave them a final form. The motion of the circle generating the cycloid was most simply conceived as uniform rolling, the slope of the tangent becoming a velocity and the curvature of the osculating circle giving an expression for acceleration: a direct link between geometry and dynamics was almost immediate.

It is well known, by the way, that seventeenth-century physics began with Galileo and ended with Newton. Let us mention, however, for future reference that its concepts were dominated by the idea of mechanism, just like curves and functions, and though Newton introduced general principles in dynamics, the notion of physical laws still remained vague. One can easily understand that attitude since, after all, only one phenomenon required a law with no known underlying mechanism: gravitation.

Finally, one must add a few words about algebra, if only because it had no direct connection with physical reality (except, of course, for yielding solutions to algebraic equations that were encountered in physics). Viete and Descartes established its basic rules, essentially as schemes of calculation, and they also introduced a convenient notation. But there were complex numbers, which had been discovered in Italy during the sixteenth century, when Tartaglia and Cardano solved for the first time algebraic equations of third and fourth order. Two more centuries were necessary before it was really understood why their method works, and complex numbers could then appear as absolutely essential.

Let us recall the underlying pattern. Taking complex numbers as granted, one denotes as usual the cubic roots of unity by $1, j$, and $j^2$, and by $(a, b, c)$ the three roots of an algebraic third-order polynomial, written, for instance, in the form $x^3 + px + q$. One easily checks that the two combinations $A = (a + bj + cj^2)^3$ and $B = (a + cj + bj^2)^3$ are invariant under a permutation of the roots $(a, b, c)$, and they can be computed explicitly in terms of the coefficients $p$ and $q$. One can then easily compute the three roots $(a, b, c)$ by solving the first-order algebraic equations resulting from the explicit values of the cubic roots of $A$ and $B$ (and the relation $a + b + c = 0$).

This event was a benchmark in the history of mathematics, because it asserted the necessity of complex numbers. There are cases when the three roots $(a, b, c)$ are real although their expressions involve inevitably the square root of a negative number. Bombelli, for instance, a student of Cardano, established the remarkable relation

$$\sqrt[3]{2 + \sqrt{-121}} = 2 + \sqrt{-1}.$$

The discovery of complex numbers shows several interesting aspects of the construction of mathematics, which will be considered in more detail later. The starting point was a problem. It was known for a long time how to solve second-order algebraic equations (in ancient Mesopotamia, Egypt, and China). The consideration of higher-order equations was suggested as a straightforward generalization, and it was perfectly natural in the perspective of exploring new territory with the experience of a familiar one. The ingredients of many great discoveries are found in this paradigm.

But we must not lose our sense of wonder! That complex numbers entered the game in that way is just a historical accident, but the place they were to hold later in the whole of mathematics and physics is a very different kind of event. A mathematician—in that case Girolamo Cardano—may invent a very clever trick and win the prize of leaving his name in the memory of mankind. But that generations of other mathematicians contemplated for centuries the incredible fecundity of the new germ and met it again and again almost everywhere is something very different. In the case of complex numbers, this extraordinary expansion was explained much later with the introduction of algebraic fields and their classification. But did that modify the situation much? Certainly not from a philosophical standpoint, because an essential fact remained unchanged: We (as human beings) may invent mathematical ideas, but they then live by themselves, and our only contribution is to be active spectators, helping to open new doors and enjoying the show.

## How Mathematics Is Made

Mathematics has something unique among all sciences: it has a corpus of results and methods, which grew over twenty-four centuries or so without any appreciable loss. The legacy of Euclid, Archimedes, or Gauss still remains valid today, even if seen in a wider frame. There would be no point, however, in trying to depict this corpus in this book as it stands now: it is too big! One may try to estimate its growth during the last two centuries by comparing it with what was acquired during the previous twenty-two centuries, until 1800, by considering either the corresponding number of pages in history books or a list of creative mathematicians[2]: the ratio is typically five to

[2] A list is given at the end of the second volume in Dieudonné (1978). It does not include any mathematician still living in 1978.

one. I shall therefore give up the prospect of sketching the history of the prolific most recent period and rather concentrate on its most appropriate aspects, when viewed from our specific standpoint.

The first aspect of interest is expressed by the question: How is mathematics made? We would have to raise the question at one place or another in this book, and the present one is convenient, since in no period of history did one dispose of richer and more instructive examples than now. It will be better also to avoid postponing this question and risk mingling it more or less with philosophical queries, since it is essentially independent of them.

## The Fecundity of Problems

Problems are the food of mathematics and the Age of Enlightenment left many unsolved problems for the perspicacity of later investigators. Some astronomical questions arising directly from Newton's theory of gravitation—such as the effect of a perturbation on planetary motions or the stability of a three-body system—nourished research for two centuries, and we owe them Hamiltonian dynamics, much of topology, the theory of chaos, and nonlinear dynamics. Newtonian dynamics also generated an eager interest in differential equations, which were investigated in much detail during the nineteenth century. Partial differential equations, particularly the linear type, had been met in the eighteenth century but they really came to the forefront of theoretical physics during the nineteenth century, with the gravitational, electric, magnetic, and electromagnetic fields. They gave rise to important concepts in differential geometry and spectral theory, and the irksome problem of the existence of their solutions was one of the main incentives for creating microlocal analysis.

Algebra drew much inspiration from two related problems, which are both easily stated: Does an arbitrary algebraic equation possess roots? Can one compute the roots explicitly by means of algebraic methods? The answer to the first question emphasized the importance of complex numbers, from which a long series of new questions and new concepts arose (there is no better source for good problems than finding the solution of one of them). The answer to the second problem (computing the roots of an equation explicitly) was not particularly interesting, but it became a magnificent example of the inner wealth of mathematics. So some people looked for the answer to this rather pointless question, just for the fun of it. I say "pointless," because if it had been found that the roots of any algebraic equation can be obtained "in terms of radicals," the result would have been more or less useless: the computation of a radical (the $n$th root of a number) is almost as difficult and approximate as the numerical solution of an equation. People got excited, however, because there was an aura around the memory of Cardano and Tartaglia and they wished to test the progress of mathematics in the meantime.

And then came this man, nay, almost a boy, Évariste Galois, who found the most deceptive answer: practically no equation of order larger than four can be solved explicitly. Normally, nothing should have exhibited more plainly the futility of this kind of mathematical research, but the query for the answer had met strong resistance from the "thing itself." No path had been found through the jungle of useless speculations and calculations, until the almost trivial idea of looking more closely at permutations of the roots arose. Symmetric combinations of roots had been known for a long time, but permutations had not been brought to the forefront before that occasion. What could then be said of their family? We know by now that their essential property is to constitute a group, and the consequences of this insight were tremendous.

The new significance of complex numbers raised a new problem in analysis: What can be said of the functions of a complex variable? It was difficult to use intuition in such a matter, but Argand's representation of complex numbers as vectors in a plane made it an easy game in geometry. Then, very soon, residue calculus revealed an unexpected practical use of these functions. This calculus, however, relied on integrals, whereas integrals had remained based on intuitive considerations until then. We have thus in this example a nice illustration of the intrinsic tightness, or coherence, of mathematics: Algebra draws attention to a field of research in analysis; research proceeds because of a correspondence between algebra and geometry; this kind of investigation, which might have been purely gratuitous, turns out to yield the most powerful tool for the explicit computation of real integrals; at the end, one arrives at a problem of elucidation and foundation, asking: What is an integral, finally?

One can mention elliptic integrals among the direct consequences of these investigations. They had been encountered in the computation of an arc of ellipse, or algebraically as some indefinite integrals involving the square root of a third- or fourth-degree polynomial. They turned out to have long-range consequences, but I wish only to mention them as an example of the enrichment of methods in mathematics. Niels Abel once tried a new trick: Rather than considering the length $y$ of some elliptic arc depending on an abscissa $x$ as a function of $x$, why not consider $x$ as a function of $y$? This "inversion of the problem" turned out to be extraordinarily fruitful and it belongs now in the toolbox of every mathematician. Many interesting problems and many fruitful notions originated later on from some inversion of a direct problem or a prior concept.

### The Case of Trigonometric Series

Trigonometric series, or Fourier series, are worth a special comment as generators of problems. Every musical sound is a sum of harmonics and, in mathematics, a sum of an infinite number of harmonics with different weights is a Fourier series. Joseph Fourier invented them in the early nineteenth century and they turned into a wonderfully versatile tool for solving partial differential equations (which were at the frontier of knowledge at that time). Fourier used them to solve important problems in heat propagation, and they were also found convenient in other problems concerning vibration and radiation (where harmonics play a natural role also).

After an idyllic period when it was found how easily a periodic function can be expanded in a Fourier series, the demon of mathematics could not miss raising the inverse problem: "What are the conditions under which a Fourier series converges, and to what kind of function will it converge?" That was opening Pandora's box. Discontinuous functions and continuous functions with no derivative sprang up from there, together with many problems involving infinite sets. Cantor's work originated from one of these problems and then much of the theory of Lebesgue's integral, harmonic analysis, and metric spaces.

The constructive process of mathematics is again clear in this example. A simple and beautiful idea, harmonics, suggests a new method for studying more thoroughly an interesting class of problems, in differential and partial differential equations. The derivation of a sine or a cosine amounts essentially to a multiplication, so that linear partial differential equations with constant coefficients turn into algebraic equations. Then the routine of standard methods, such as generalizing, inverting, elucidating the foundations and the boundaries of a theory, led in very different directions and raised new fascinating problems. What may be kept in mind from this example for our future reflections is certainly the intimate relation of problems arising immediately from physics with some of the most abstract concepts in pure mathematics. Is it because they belong to the same world?

### Structures

We saw that abstraction is a basic process in the brain. Another essential feature of such processes is the association of ideas, which is not yet really understood by cognition sciences. It goes also under other names such as analogy, correspondence, similitude, or equivalence. Mathematicians have brought their recognition to the level of an art (see Polya 1945, 1954a, b for illuminating examples). We mentioned the group structure already, and Hilbert emphasized the importance of the idea of structure; Bourbaki systematized it in his *Elements of Mathematics*, and a list of the structures he used would be tantamount to a Table of Contents of that impressive work, much too long for us to give it here (the elucidation and the choice of the most convenient structures, by the way, were the topic of much discussion in the Bourbaki team).

The use of structures has become one of the main tools in mathematics. Two features seem to dominate their choice: economy and efficiency. When a known structure is recognized in a given mathematical object, it immediately provides a great economy of thought, since it brings with it automatically a number of concepts, methods, and useful lemmas. This convenient method has certainly played a great role in the acceleration of math during the last two centuries. Quite a few structures, however, were proposed, investigated for some time, and finally left over if not forgotten, because they were too general to be really fruitful. The function spaces that are used nowadays, for instance, are more specific than the very general ones that were proposed during the first half of the twentieth century. This recognition of the efficiency of structures brings forward the question of their existence: Why do some of them appear as unavoidable keystones in the architecture of mathematics? This question is too close to the still more basic one, "What is mathematics?" to be considered presently, and we had better leave it aside for some time.

Axiomatism is in some sense the search for a minimal structure of structures, and the philosophical questions it raises are not essentially different. It brings up, however, new *problems*, which belong mostly to metamathematics, i.e. a theory of mathematics that was particularly impelled by Hilbert. The consistency and independence of a set of axioms are typical of these problems, as well as the complete character of the resulting branch of mathematics (a theory is complete if every proposition A in it is such that either A or non-A is a theorem). As is well known, Gödel has shown that formal arithmetic is not categorical, but axiomatism nonetheless remains essential in the practice of mathematics.

There are other important aspects of the association of ideas in the methods of mathematics. The notions of isomorphism, homomorphism, and other "morphisms" are examples of a correspondence between different objects having a common structure. Equivalence—a strong form of analogy—is a universal type of structure implying the existence of classes of equivalence and providing one of the most powerful techniques for constructing new objects. Structures can also be used to specify a definite type of theory, and the best-known example is Klein's classification of geometries according to their invariance groups.

Last but not least, one can use models. A model consists essentially of an example or a formulation of an abstract theory using the framework of a simpler one. Complex numbers, for instance, can be defined in the framework of real numbers as pairs of real numbers with suitable rules for addition and multiplication. Rational numbers can be obtained as an equivalence class of pairs of integers $(p, q)$ with the equivalence $(p, q) = (p', q')$ if $p \cdots q' = p' \cdots q$. Real numbers have also been constructed as models on the set of rational numbers, so that integers became a building block for every kind of number. When a model provides only an example and not a universal representation of a general theory, it proves at least that this theory is not empty.

I am sure that I left out quite a few significant tools in the mathematicians' toolbox, but the conclusion is clear enough: No philosophical consideration is necessary when one is doing mathematics, no more than when playing any sort of game. Conversely, this autonomy gives no hint toward explaining why mathematics developed into a gigantic construction, so fecund, so consistent throughout, and with such a mysterious ability to fit reality, even now when reality is no longer intuitive and has become nonclassical.

## MATHEMATICS AND REALITY

We saw mathematics come forth initially from an abstraction of reality. There remained little of that original relation at the end of the nineteenth century, however, after the search for more inherent foundations. It was then fashionable to declare that no obvious—or even remote—relation with external reality should constrain the construction of mathematical science. The main steps of this enlargement of the global frame are well known. There was the analogy between linear algebra and $n$-dimensional geometry; after having been initially a source of models, it rapidly became systematic, and spaces of arbitrary dimension were turned gradually into familiar and almost intuitive concepts. Non-Euclidean geometries had a much more drastic effect. Although every book in history mentions Gauss's attempt at checking the sum of angles in a triangle by making geodesic measurements, the word "space" received two noteworthy different meanings soon after Lobatchevsky and Riemann: the ordinary, real, space was left to physics, while mathematics collected more and more nonfigurative objects with the same name (though calling them "spaces" was only the far-off outcome of a long chain of successive abstractions).

The second half of the nineteenth century saw an invasion of pathologies in analysis, beginning with the discovery of continuous functions with no derivatives. One often quotes Hermite's reaction, who saw them (with a touch of humor) as a "lamentable sore" and who "turned away from them with horror and fear"; but as noted by Bourbaki (1960) "So many monsters of that sort have been encountered since one century that we feel somewhat *blasé* and only an accumulation of the wildest unnatural characters can still surprise us." Poincaré, however, wondered earnestly "Why can it be that intuition deceives us so much?"

The link of mathematics with reality was slowly broken. When inaugurating the axiomatization of algebra, Hankel proposed in 1867 a view of mathematics as "a pure, exclusively intellectual theory of forms, whose object is not a combination of quantities or of the numbers, which are their images, but of the things belonging to thought, to which it may happen that real objects or real relations correspond though such a correspondence is unnecessary." Cantor advocated in 1883 a "free mathematics," claiming that "the development of mathematics is entirely free and its concepts are constrained only by the necessity of noncontradiction and the condition of being coordinated with previously introduced concepts through precise definitions." According to a well-known story, Hilbert expressed the same idea by saying that the words "point," "straight line," and "plane" could be replaced by "table," "chair," and "beer mug" without changing anything in geometry.[3]

Henri Poincaré (1866–1912) was the last influential mathematician to support the relevance of reality in the existence of mathematics. Such was in fact his style in research, and he had always been much involved in theoretical physics and geometry, where he found a large part of his inspiration. He was also much interested in the process of mathematical invention and he could not accept that the creation of mathematics would be reduced to a strict matter of logic: it relies too much on intuition, and intuition stems ultimately from reality. He was therefore unwilling to reduce a choice of axioms to a pure convention and he wanted to preserve some "meaning" in the axioms, if only by keeping memory of their historical and psychological origin. Regarding the question of foundations, he expected that the existence of integers should be a sufficient ground for establishing the edifice of mathematics and he saw that position as a sensible middle way between

---

[3] These quotations are borrowed from Bourbaki 1960, 34.

Cantor's "free mathematics" and the ultimate relevance of reality, or at least of meaning.

## The "Crisis" of Mathematics

Then came the famous "crisis" in the foundation of mathematics on which so much has been written. It is considered essential by most philosophers of mathematics and one must certainly take it into account, but it has somewhat less significance in physism, which is advocated in this book. None of the main trends in that period, formalism, logicism, or intuitionism, had any direct relation with the present approach—except for Poincaré's general outlook—and their discussion will find a better place later on. The case of intuitionism warrants a mention, however, since its proposal by Brouwer extended some of Poincaré's suggestions, and one might wonder whether it could be an ancestor of physism in one way or another. This is not so, however, and the reasons for this rejection are twofold. The first one is already well known: strict intuitionism—in which everything is built on the integers—disallows too large a part of mathematics, including some that is now found necessary in physics; it is therefore inconsistent with physism. A second reason, which will be explained later, is the place of integers in the foundation of physics, which is certainly more problematic than in mathematics. The main point in that case is that intuitionism relies essentially on classical reality, whereas reality is fundamentally *nonclassical*, and the background has therefore drastically changed.

The crisis must be briefly discussed nonetheless, although not for its own sake. Reality has changed meanwhile, if I may say so, or rather our knowledge of it has. There was also some sort of crisis in physics with the advent of relativity and quantum mechanics, and the comparison with the case of mathematics will be instructive. I will not introduce this parallel for purely philosophical reasons, but rather for historical and psychological ones, because it explains why practically nobody suggested any reference to physical reality in the philosophy of mathematics for almost a century. This old idea meanwhile became so old fashioned that one must understand why it was lost so completely. The history of ideas provides the necessary explanation, and the times at which the two crises occurred in mathematics and physics turn out to be illuminating, together with their respective durations. This is why the present account of the mathematical crisis will be limited to a few dated milestones, with little comment and no intention whatsoever of getting to the bottom of the difficult underlying questions, and it will be followed by a similar discussion of the case of physics.

### Naive Sets and Their Paradoxes

The elucidation of analysis, including integrals, convergence, and a first understanding of real numbers, had revealed the pathologies we mentioned already and had made intuition suspect. It had become clear that mathematics is primarily concerned with relations and not with quantities and objects, the general definition of a relation or a structure requiring a foundation on precise axioms. Much effort toward a proper axiomatic therefore took place in the last quarter of the nineteenth century. Frege's *Begriffshrift* (1879) was the main landmark in the case of mathematical logic. It introduced many fundamental notions, among them truth functions and propositional calculus; propositions were decomposed into function and argument instead of subject and predicate; a new system of logic carried out inferences according only on the form of propositions and not their supposed meaning. Peano's *Principles of Arithmetic* (1889) and *Formulaire* (1895) dealt with the axioms of arithmetic and Hilbert's *Foundations of Geometry* (1899) gave a complete list of axioms for elementary geometry.

A common feature of all these attempts was a free use of the notion of set, which had been made central by Cantor.

Everybody thought that an expression such that "the set of . . . such that $P$" ($P$ denoting a property) properly defined an ensemble of mathematical objects, which could be investigated through an analysis of the consequences of $P$. But suspicions arose at the end of the century. Burali-Forti noticed in 1897 that the infinite collection of ordinal numbers could not be a set to which Cantor's methods would apply. Cantor himself, in 1899, remarked that the same is true of the collection of cardinal numbers and of "the set of all sets." Russell, in a letter to Frege dated 1902, called his attention to the fact that "the set of all sets that are not elements of themselves" is a self-contradictory notion. Richard (1905) showed that some finite sets have no sensible definition and, for instance, "the set of integers that are defined by less than fourteen English words" is certainly finite, but "the smallest integer that cannot be defined in less than fourteen English words" is a self-contradictory notion, since its definition requires only thirteen words (the example can also be translated to information bits on a definite Turing machine). There was something devious in these paradoxes, which often looked remote from the center of mathematics, but they pointed anyway to the necessity of an axiomatization of set theory.

## The Foundations of Set Theory

Russell reacted to the paradoxes in 1908 by introducing his theory of types, but we shall concentrate on the work of Hilbert's disciples and particularly Zermelo, who proposed in the same year a system of seven axioms for set theory. Axiom III (the axiom of separation) removed the paradoxes by introducing a simple condition in the definition of a set "such that $P$." It introduced the condition that the property $P(x)$ be defined on all the elements $x$ of a previously given set $E$ and then the collection of all the elements for which the property $P(x)$ is true is a proper subset of $E$. It was observed, however, soon after that the property $P$ must itself be properly defined in an axiomatic way and, consequently, the axioms of formal logic must stand prior to those of set theory.

There was also an Axiom VI, the axiom of choice, which started many controversies. It is concerned with a set $E$ whose elements are disjoint nonempty sets $X$. It then states the existence of a set $M$ sharing a unique element $x$ with every set $X$. Zermelo had already introduced this axiom four years earlier to prove an important theorem, namely, that every set can be well ordered. This consequence is sometimes considered as another axiom, equivalent to the axiom of choice, and it sets up a cornerstone without which a large part of modern analysis would collapse. Its significance in theoretical physics will be considered later, but the theory of Hilbert spaces and quantum mechanics certainly need it; it plays therefore some part in our understanding of physical reality.

The axiom of choice is very troublesome anyway, because it assumes that an infinite and often uncountable number of operations (choices) is performed. What does that mean? Some critics were very severe and Brouwer's intuitionism originated in a strong reaction against the new proposals. We cannot discuss Brouwer's ideas in any detail, if only because they link up psychological considerations with logical ones, a privileged role of integers in the foundation of mathematics with strong restrictions on the principle of the excluded middle, together with subtle mathematical developments. A famous theorem by Paul Cohen (1963) is certainly more rewarding, since it implies that the axiom of choice is independent of the other axioms of set theory: one may assume it or not, and obtain in each case an equally consistent set theory.

## Hilbert's Program

At the center of the controversy around the axiom of choice was the question of existence in mathematics. This is particularly clear in a well-known exchange of letters between Baire, Borel, Hadamard, and Lebesgue (Baire et al. 1905). Under what conditions can one assume that a mathematical object exists? Should it be constructed explicitly, and, if so, what are the restrictions on the construction processes? The connection of these questions with realism is particularly transparent in Brouwer's intuitionism, so much so that the only kind of functions he allowed had to be continuous: the relation to classical reality is obvious. Hilbert's approach was drastically different and, had it been successful, it would have implied the existence, in itself, of a Platonic reality, independent of physical reality, "for the greatest glory of the human mind."

In a famous conference at the International Congress of Mathematics, in 1900, Hilbert had already proposed a new principle: whereas the noncontradiction of a concept made it only "possible" in the framework of traditional logic, he considered noncontradiction as ensuring the existence of the (axiomatically defined) concept. Poincaré had immediately noticed a key difficulty in this assumption, since it implies apparently that one must prove a priori the noncontradiction of a mathematical theory before being permitted to develop it legitimately. Hilbert then spent thirty years trying to meet the challenge. He concentrated first on formalized arithmetic and later extended his investigations to the theory of real numbers and set theory. Several questions were then raised, the main ones being the following.

- *Noncontradiction.* Can one assert that every proposition (which is written in agreement with the formal grammar of the relevant theory) is necessarily either true or not true? This means in practice that there exists in principle a proof (involving a finite number of steps), relying on the basic axioms and concluding the veracity of the proposition or its negation. The analysis of this problem led, by the way, to a great progress in the theory of proof.
- *Independence.* The propositions $\{A_i\}$ in a given system (belonging, for instance, to a selection of some axioms) are independent if no proposition $A_i$ is a theorem in the system where the axioms are the remaining propositions $\{A_j\}, j \neq i$.

- *Decision.* A theory is decisive (or categorical) if every proposition $T$ belonging to it is a theorem, or non-$T$ is a theorem. This problem had a long history later in algorithmic theory. It asks whether a universal algorithm can decide whether a program on a Turing machine will halt after a finite number of steps (a Turing machine is essentially an ordinary computer, except that it can run in principle for an infinite number of calculation steps). The answer was found to be negative.

The program failed ultimately with Gödel's theorem (1931), which marked in some sense the turn of the tide in the great crisis. Roughly speaking, this theorem states the impossibility of proving the noncontradiction of arithmetic (or of any theory containing arithmetic as a subtheory). Its initial proof was difficult, but it is often stated more simply now in the framework of information theory (see, for instance, Chaitin 1982). Finally, Hilbert's grand purpose had failed and it left the philosophy of mathematics an orphan. The theory and practice of computation has since been the main beneficiary of the powerful techniques that were developed along the way: reality, or rather pragmatism, has been the winner at the end of the day.

## A Parallel with Physics

There are remarkable analogies between the histories of mathematics and physics (particularly quantum physics) during the twentieth century. I intend only to point them out in a rather rough way presently, however, because the second part of this book will deal in more detail with the questions of quantum mechanics. Both sciences have known a "crisis," as many people say; but it was not a crisis of science, since science continued to grow at a tremendous rate, essentially unaffected. It was each time a crisis in the philosophy of knowledge, and not one of knowledge.

We just saw that the core of the mathematical crisis was the question of existence, and there were some indications that it had some relation to the question of reality. In physics, the nature of reality was directly at the center of the concerns. We left it in the previous chapter in its classical (intuitive) state and we noticed the corresponding main characteristics, namely, uniqueness, location in ordinary space and time, continuity, separation of phenomena, causality, and what was called a sense of reality, which will now be expressed better as a clear-cut distinction between what stands as real and what is only virtual.

Quantum mechanics struck a blow at all these qualities, just as mathematics separated objects and quantities. Continuity was the first victim in both cases. In physics, quanta marked the end of continuous classicality, already in 1900 with Planck. It was much more damaged when Heisenberg's uncertainty relations were discovered in 1927, since it was then found impossible to *think* of the motion and the trajectory of a particle in space: its location in space and time had become completely fuzzy. Causality had already been disparaged in 1926 with Born's probabilistic interpretation of wave functions. The nonseparation of phenomena, already stated by Schrödinger in 1926 and despite the doubts of Einstein, Podolsky, and Rosen in 1935, has been unequivocally confirmed by recent beautiful experiments on the so-called "entangled states." The uniqueness or nonuniqueness of reality is a subtler matter, which has many facets on which I do not intend to dwell right now; its discussion began with Bohr's "complementary principle" back in 1928 and it is definitely nontrivial. What is left then? Is it a clean difference between reality and virtuality? But it certainly did not remain clean at all after the introduction of "virtual processes" around 1950!

Wolfgang Pauli stressed the fact that almost every category of reason (according to Kant) is disparaged by quantum mechanics, so that the philosophy of knowledge was deprived of its safest foundations. These dire conceptual difficulties

45

were made still more acute rather than alleviated by the fact that empirical reality, at large scales, remains perfectly classical (with the meaning we gave to that word), and Bohr's evocation of a rather vague "correspondence principle" relating it with the realm of quanta did not help much. A clever attempt by John Bell, in 1964, to recover classicality at a lower level unfortunately failed twenty years later. One can therefore conclude in a nutshell that the problems of reality appeared by and large much worse than the relatively unassuming problems of mathematical existence, and one understands easily why physics was not called to the help of mathematics.

The analogy between the histories of mathematics and physics extends to the occurrence of paradoxes, the most famous ones being the Einstein-Podolsky-Rosen paradox and the Schrödinger's cat paradox. Paradoxes occur when one applies common sense to some tricky quantum situation and, since common sense is born from the regularities of our surroundings, it turns out to deceive us often. Logic is easily tricked. Worse than that, some careful analyses of the meaning of experiments by Bohr were too often summarized hastily as meaning that quantum physics is not objective. The fact that a definite experiment can measure one quantity and not another, position and not momentum, an electric field and not a photon, was misinterpreted as an action of the observer on physical reality. No wonder then that quantum physics has been suspected or ignored by many philosophers and laymen, in spite of results that outperformed the total amount of physical knowledge during the three previous centuries.

Philosophy has remained more or less silent on these difficult questions, except for expressing worries. When Husserl spoke of a crisis of science, in 1935, he referred only to mathematics, and his considerations on physics were shallow. His phenomenology is definitely incompatible with quantum science, as are Wittgenstein's states of affairs. Some contemporary philosophers are fortunately aware of these problems, but no general philosophy of knowledge is yet in view, although, if the problems of knowledge in mathematics and physics are effectively linked, one certainly needs a wider framework.

Significant progress in the understanding of quantum theory occurred during the last decades of the twentieth century. Paradoxes, for instance, have now completely disappeared, and the emergence of the classical features of reality from the quantum laws is well understood. Some controversies remain, unavoidably, because people's minds change less rapidly than knowledge, but the general trend is unmistakable. I will strongly rely on it in this book, and it will be explained and discussed, together with its significance for the relation between mathematics and reality. Since this chapter was mainly devoted to history, one may conclude it with what I consider a significant remark about the timing of the two crises, in mathematics and physics. In mathematics, it began in 1902, reached the turn of the tide in 1931, and was essentially allayed in 1963. In physics, it began in 1900, went through an apex in the late 1920s, and maybe (I would say probably) is presently reaching still waters. This time may therefore be the right one for drawing the lessons of the turmoil we have passed through and reconsidering earnestly whether mathematics and reality have something deep in common.

# PART TWO

REALITY AND THE QUANTUM WORLD

# First Encounter with the Quantum World

There are certainly good reasons why somebody interested in the philosophy of mathematics might be unfamiliar with physics, and particularly quantum physics. It looks at first sight very remote from anything like the foundations of mathematics and, if one is interested in philosophy, the end of the previous chapter mentioned difficulties with reality that were not very encouraging. The aim of this book, however, is to show that the two questions, the philosophy of mathematics and quantum reality, are very closely related, so that we must resign ourselves to entering the quantum world at one time or another. This world is not a jungle, however; it is not a labyrinth, but a huge temple with many degrees of initiation, and I will try to make the entry as easy as possible, although I do not pledge myself that all the steps will always be comfortable.

An easy access should avoid encountering many unfamiliar ideas at the same time, and I think that the Feynman approach to quantum mechanics was most convenient for that purpose. Its mathematical tools are no impediment and it shows immediately why quantum science privileges possibility over actuality and the interplay of many virtual alternatives over a unique sharp reality, and why nonclassicality stands at the heart of the kind of realism we need as a framework. Some readers will certainly wonder why they are asked to abide by the strange quantum rules and what facts make them unavoidable. I can only say that I am much in sympathy with their pragmatic request, but that is the topic of many other books and, at any rate, all physicists agree that quantum mechanics fits every known fact, including many that were obtained to confirm the strangest consequences of the theory. There are, moreover, more and more indications that it is a self-consistent theory, which makes it a good partner for a final comparison with mathematics and a conergence of their problems. But we are still at the beginning and we must proceed.

## A Game on a Chessboard

Let us imagine a large horizontal chessboard with many little squares. A particle can move freely on it. Two kinds of games can be played on it, with different rules, the classical ones and the quantum ones. In the classical game, the particle moves freely on the board, always along a straight line with a constant velocity. The quantum game is less simple. The particle is first placed on a square A (any one): this is the preparation process. Then it moves but we don't see it. There are two different methods for discovering where it stands. The first method is a "position measurement." It consists in throwing a net on the board at a definite time T, each mesh of the net covering exactly one square of the board; the net traps the particle and one can check the mesh in which it has been caught. Another method consists in using a detector. In that case, one side of the board is sticky and, if the particle hits this detecting side, it is glued and stays at the place where it was caught until one looks at the detector and registers where the particle stands. These measurement processes may look tricky, but they are part of the game and we shall come back to their meaning later.

We can devise various experiments by disposing obstacles on the board. A square, for instance, can scatter the particle, which cannot enter it. Rather than a unique square, we can use a set of squares showing any shape. A particularly interesting device consists in putting a wall along a row of the chessboard somewhere in the middle, leaving only two open squares through which the particle can go (see figure 4.1). This will be called a Young device, in memory of Thomas Young, who used a similar system for observing the interference of light long ago.

Then we perform a series of experiments. Every experiment is made with a definite obstacle (or no obstacle at all) and a definite measurement device (using either the net or a detector). One performs many trials under exactly identical

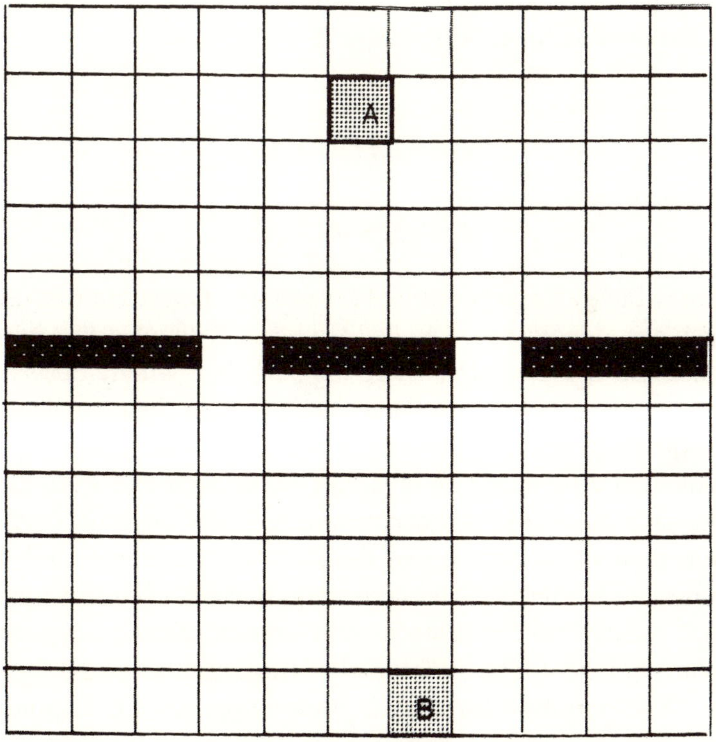

Figure 4.1. An elementary approach to quantum mechanics.
A particle starting from square *A* moves on a chessboard. An obstacle consisting of a wall with two apertures realizes the conditions of an interference experiment.

conditions. The particle starts always from the same point $A$, and one looks in which mesh of the net it stands at some time $T$ (in the case of a position measurement), or where it is glued on the detector. A very important experimental result is that *the results are randomly distributed.* In the case of a position measurement, for instance, the particle is found in one mesh or another, and one may perform a long series of individual trials, make a statistic of the results, and derive from it a probability for the particle to be found in each square (always after the same time $T$). We will not try to draw premature conclusions, however, such as asserting that this quantum game must obey purely probabilistic laws; we only measure the probabilities, without deciding whether hidden causes, if any, are responsible for randomness.

## FEYNMAN HISTORIES

### *In Search of a Rule*

We now have a large amount of probability data at our disposal, which we denote by $P(A, B; E)$ where $A$ denotes the initial square and $B$ the square where the particle is finally observed when we use a net. When we use a detector, $B$ denotes the side of a square on the boundary of the board. $E$ stands for the experimental conditions: which obstacles are present and what kind of measurement is performed; it also includes the time $T$ when one uses the net. Notice that the size of the squares (and the meshes) also enters in $E$, and we could make it arbitrarily small, which means that the obstacles can have in practice any shape and arrangement. This is definitely a huge amount of data!

We then look for a basic underlying law, i.e., a unique mathematical rule generating the data. Perhaps it could be written down in several different ways, but we ask only for one of them, feeling sure that mathematics will help us afterward to find other equivalent ones, if any. The most convenient rule in that case turns out to be one discovered by Richard Feynman in the late 1940s, which will be now stated.

## Nonclassical Behavior

A classical particle moves along a straight line with a definite velocity, but we did not say anything about the initial velocity; we just gave a starting point. Could it be that the randomness we contemplate means only that this velocity is random? We would then only have to find its probability law. But this idea does not work. The Young experiment is the cleanest test for that. Suppose that the initial point *A* is on one side of the wall, far enough from it, and the detector is parallel to the wall (along one side of the board opposite to *A*), also far enough. If the particle moved classically, the data would be the same as when an awkward shooter aims at the wall: the impact points would be concentrated along the two directions joining *A* and the two holes. But that is not what is observed: One finds a nice pattern of fringes on the detector, similar to the interference pattern that was observed by Young in the case of light.

That's fine, you might say, it means that the quantum phenomena involve waves; waves suppose a wave equation and we have only to find it! This suggestion is good, and it would bring us to de Broglie's waves, wave functions, and the Schrödinger wave equation. And what then? Then we would be in a conceptual mess where all the difficulties would fall on us together as they did on the founders of quantum mechanics. We want to avoid that, which is why we turn to Feynman histories. They are equivalent to the wave approach from a mathematical standpoint, but they split the concepts much more nicely and we shall therefore use them.

## Feynman Histories

The shooting example has shown us that the particle does not move classically. Since it can reach some places on the detector that could not be hit through classical motion, this might mean that the particle trajectory is more complicated than a straight line; it curves, it wiggles, perhaps; what is it? But why do we speak of "the" trajectory? Feynman's brilliant idea was to imagine that the particle has no special trajectory, not even an arbitrary one, but has in some sense all the conceivable ones: quantum mechanics is not only a probabilistic theory but the laws of a world where all the possibilities exist together! "Together" means that they do not stand as alternatives but contribute equally to the final result, and none of them is dispensable.

Let us put some math in that. We define a Feynman history as an arbitrary motion of a point $x(t)$ on the board. To what extent arbitrary? Let us say that a Brownian motion is a good example, although we do not assume an underlying probability determining its possible choice (see figure 4.2). (By the way, we leave the question of continuity or wider conditions to highbrow mathematicians and think of a continuous differentiable motion for simplicity; more generally, every mathematical nicety will be left out in the following.) Every motion is allowed, and we dispose accordingly of a huge set of functions $x(t)$ with velocities $v(t)$. When there is an obstacle on the board, no Feynman trajectory can penetrate it (the case when a force is exerted by the obstacle at a distance will be considered later on). When there is a naive detector such as the ones we considered, the motion stops at the point where it hits the detector, if it does.

The simplest case of an experiment occurs when there is no obstacle and no detector, just a net falling at time $T$. We are then interested in the probability $P(A, B)$ of finding the free particle in square $B$ at time $t$ when it started from square $A$ at time $0$. Let us suppose that the squares are small enough that $P(A, B)$ varies slowly with their positions and let us concentrate on the motions starting from a definite point $a$ in square $A$ and reaching a point $b$ in square $B$ at time $T$: i.e., $x(0) = a$, $x(T) = b$.

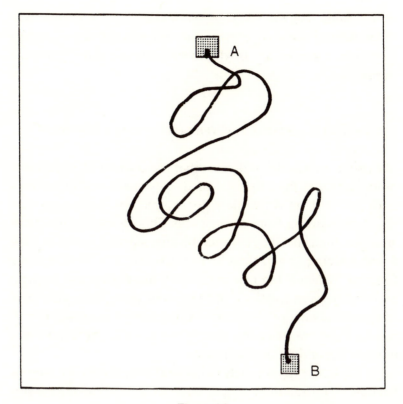

Figure 4.2.

Feynman's formulation of quantum mechanics then holds in three formulas, which we write as

$$P(A, B) = l^2 \, | \, \text{amplitude}(a, b) \, |^2, \tag{1}$$

$$\text{amplitude } (a, b) = \int \exp\!\left( \frac{i}{\hbar} \, \text{action} \right) d(\text{histories}), \tag{2}$$

$$\text{action} = \int_0^T L(x(t), v(t)) dt. \tag{3}$$

I suppose some mathematically literate readers will ask for more precision whereas others may prefer something more intuitive, and I propose therefore to examine these formulas in two steps. We shall first consider the effective mathematical content of these equations and, in a second step, we shall try to use a simple toy model to give them a more palatable expression.

## The Mathematical Content of Feynman Histories*

(As usual, an asterisk indicates that the content of a section is more technical than the rest of the chapter to which it pertains.) It will be convenient to explain the basic formulas (1)–(3) by starting from the last one. This equation (3) defines an action for a specific Feynman history, starting from $a$ at time 0 and arriving at $b$ at time $T$. The function $L$ is the Lagrange function, which is familiar in classical mechanics and is given by the difference $L = K - V$ of the kinetic energy $K = \frac{1}{2}mv^2$ and the potential energy $V$, which is a function of the position $x$ ($L$ differs therefore from the energy $E = K + V$). The action is a number depending on the history.

Equation (2) is much richer and constitutes the core of Feynman's formulation of quantum mechanics. We must therefore decompose it piece by piece. Let us look first at the exponential, which refers to an individual history. The exponent involves the action of the history at hand (beginning at $a$ and ending at $b$) and we just saw how this quantity is defined. An action has a physical dimension, which means that its numerical value depends on the choice of the units for length, time, and mass.[1] We also see a quantity $\hbar$ in the denominator, which is conventionally Planck's constant $h$ divided by $2\pi$. This constant controls everything in the quantum world, and its presence indicates therefore that we are referring to that world. Planck's constant has the same physical dimension as an action, so that the quotient $a = action/\hbar$ is dimensionless, i.e., a pure number, which we shall call the *phase* of the history.

---

[1] The physical dimension of an action is $ML^2T^{-1}$.

The last factor in the exponent is the imaginary number $i = \sqrt{-1}$. The occurrence of complex numbers in physics has always been a subject of questions, but it should not worry us too much. It appears unavoidably in intermediate expressions from which actual physical quantities can be computed, but of course never in these quantities themselves. The exponential itself $z = \exp(i\alpha)$ is a complex number of unit modulus, which can be represented in the complex plane by a vector of unit length making an angle $\alpha$ with the real axis. We see therefore that every history makes a complex number $z$ available for later calculation, and only that; all these numbers have the same modulus, the same weight so to say, and none is more important than another. In other words, no history is fundamentally more important than another. Their only difference consists in their phase (but that can do much).

This equation apparently involves an integral, but it is a rather special one. It may be remembered that the integration sign ($\int$) is a slight deformation of the letter "S," the initial of "sum," and was introduced by Leibniz to mean a continuous summation. It is followed by the quantity to be summed, and the continuous index on which summation is made is preceded by the letter $d$: for instance, a continuous summation over a variable $x$ is denoted by $dx$. Here, we have $d(histories)$, which means that the sum is made over all the histories beginning at $a$ and ending at $b$.

Although we agreed to avoid questions of rigor, a word of caution is necessary here. The set of histories is enormous and the usual definition of an integral does not apply to a sum over histories. It can be given a meaning through a limiting process in simple cases: the case of a finite number of nonrelativistic particles interacting through potentials (as in our example) or the case of relativistic particles interacting through classical fields. Physics, however, needs more than that. In quantum field theory, the quantity $x(t)$ indicating the possible position of a particle is replaced by the possible values of a field (e.g., the electric field) at every point of space-time. One does not know then how to rigorously relate the corresponding Feynman summation to the conventional axioms of functional analysis, although one knows perfectly well, empirically, how to evaluate Feynman sums and manipulate them. One is, however, treading in still uncharted no-math land, while physics shows convincingly on the other hand that this expression of quantum laws is the most general and most efficient one. Nonetheless, this peculiar situation should not worry us too much and we may remember for instance that the Heaviside calculus and Dirac's delta function had shown their efficiency long before the advent of distribution theory, which gave them a reliable foundation. Quasiempiricism, a philosophy of mathematics to which we will come back later, also has nothing against this way of doing math. We had better close, therefore, this digression on matters of principle.

Equation (2) also contains a definition: the sum on the right-hand side defines a quantity that is called the "amplitude." This is short for the expression "probability amplitude for the particle to go from $a$, starting at time 0, to $b$ at time $T$" and it is clearly a complex number. The meaning of this probability amplitude is made clear in equation (1), which gives the probability for going from the little square $A$ (containing point $a$) to square $B$ (containing point $b$). The factor $l^2$ in the right-hand side is the area of a square (with side $l$), so that the square of the amplitude appears as a probability density in space, as used in probability calculus. This is something we understand perfectly well intuitively: the probability of receiving a raindrop on a paving stone is the product of the probability density for receiving drops at that place, times the area of the stone. Equation (1) says finally how one can compute the data $P(A, B)$ from the basic concepts of quantum mechanics, which are Feynman histories and Feynman sums in the present approach.

We may be brief in the application of these rules. In the case of Young's device, the summation will have to be made over the histories crossing either one of the two holes in the wall and ending up at some place $b$ on the detector. The Feynman sum (2) then involves two quantities: the contribution from the histories crossing one hole and from the histories crossing the other hole. The probability in equation (1) is then the square of the sum of two complex numbers, and an explicit calculation shows that one finally obtains an interference pattern on the detector.

When there is an external potential $V$, it should be included in the Lagrange function. The extension of equations (1)–(3) to three-dimensional space rather than the two-dimensional chessboard is obvious. When one is dealing with a system of $n$ interacting particles rather than one, the points $a$ and $b$ should be considered as two points in the configuration space of this set of particles and this is conceptually trivial, even if the calculations may not be so.

## THE INTUITIVE CONTENT OF FEYNMAN HISTORIES

We may now look at the basic formulas in a more naive way, somewhat like playing a computer game. Suppose we replace the particle by a red toy car, without a motor though able to move on the board without friction. It is initially put on square $A$ and we let it go. We surely cannot see it moving, because it would mean that it has a definite trajectory, and we found that this is impossible. In our game, the true red car is immediately replaced after the start by a multitude of identical pink cars, which we may call its clones or its virtual representatives, each one of them moving in a completely arbitrary way, i.e., undergoing a Feynman history. When a position measurement is performed at time $T$, all the clones materialize suddenly into a unique red car, which is trapped in a mesh of the net, over some square $B$. The problem is again to compute the corresponding probability.

Suppose that each clone carries a dial, like the one indicating the velocity on a real car, except that the pointer of this one can turn around the dial. There is also something peculiar about this pointer. Its position at a given time does not indicate the velocity of the clone at that time, but the rotation velocity of the pointer is proportional to the kinetic energy: it turns four times more rapidly when the velocity increases by a factor of two.[2]

In order to compute the probability $P(A, B)$, one considers all the clones that are in square $B$ at time $T$. Because of the infinite number of possible histories, one may think of a very large though finite number of clones starting from square $A$ and moving arbitrarily on the board: Each clone pointer can be considered as a little vector, always with the same length because of the identity of clones (this equality in the length means that every history has an equal importance in the final result!). We then add all these vectors, joining the pointers end to end to obtain their sum, which is the amplitude for going from square $A$ to square $B$ in a time $T$. The probability $P(A, B)$ is then simply the square of the length of that sum! Not quite so, however, because there is a normalization factor involving the number of clones in the sample, the size of their dials, and the area of the squares, but this is easily dealt with by imposing that the sums of $P(A, B)$ over all the squares $B$ is equal to 1. Isn't it easy?

## WAVE FUNCTIONS

The wave function at time $T$ at a point $x$ on the board is simply defined as the amplitude in the square containing $x$ when all the squares become infinitely small. We need not worry about normalization, because it is easily dealt with. The wave function

---

[2] More precisely, the rotation velocity is the quotient of the kinetic energy by $\hbar$, i.e., the rate of change of the action in units of $\hbar$ for a free particle, according to equation (3). The angular position of the pointer on the dial at a given time is therefore the phase of the exponent in equation (2). We thus have a nice intuitive representation of this phase.

$\psi(x, T)$ is a complex number. When we are dealing with several particles, the wave function $\psi(x_1, x_2, x_3, \ldots, T)$ depends on very many variables. The construction of histories shows easily that a clone in square $B$ at time $T$ had to be in some square $C$ at a previous time $t$, so that the amplitude for going from $A$ to $B$ in a time $T$ is the sum over all the squares $C$ of the amplitudes for going from $A$ to $C$ in a time $t$, multiplied by the amplitude for going from $C$ to $B$ in a time $T - t$.

This is very interesting, because it shows that the time evolution of a wave function is *linear*. We may add two different wave functions to start with and, at any time later, the resulting wave function will still be the same sum. The sum of two wave functions is again a wave function! The linearity of quantum mechanics is an extraordinarily simple structure inside an otherwise very elaborate one. It implies a *universal occurrence of interferences* and this point warrants a few comments.

We know, for instance, from wave optics how interferences occur in a Young device. The same is true in quantum mechanics. The propagation of a wave inside a random medium is also an interesting example. Suppose for instance that an optical wave arrives from a vacuum and penetrates into such a medium (water, for instance). The electric field of the wave produces an oscillating electric dipole in each water molecule. Each dipole emits a wavelet, so that the field inside the medium is apparently a very complicated interfering sum of all these wavelets together with the initial wave (which is damped, however, because of the loss in energy resulting from the emission of wavelets). The calculation of the resulting effect is nontrivial, but the result is well known: all the fields add up to produce simply a refracted wave, which propagates in the medium with a velocity involving the refractive index of that medium! This is a beautiful example of how interferences can be sometimes constructive and sometimes destructive: they construct the refracted wave going along the refracted ray and they destroy every trace of the wavelets in a different direction. This is a very deep and inspiring result, which we will meet again.

Coming back to our example of the chessboard, we may notice that there is no necessity for letting the particle start from a definite square $A$. That was only the consequence of a particular preparation process, and another process would produce some initial wave function, depending on the nature of the devices preparing the particle (accelerator, radioactive source, laser, collision of an incoming particle with a target, and whatever one may think of occurring in a laboratory or in nature).

One can use the method of Feynman histories to get an equation for the evolution of the wave function. The calculation has some analogy with the construction of a refracted wave from the interference of wavelets: the clones, when interpreted mathematically, build up wavelets through their tiny amplitudes, and the wavelets build up the wave function. The resulting equation is the Schrödinger equation, which was of course discovered long before Feynman's histories. It will not be written down, because the three equations we wrote in this chapter will be enough for our philosophical purpose.

## CHAPTER FIVE

# Quantity and Reality

In this chapter and the next one, we consider two questions concerning physical reality: localization in space, which we may remember as an essential feature of reason according to Kant, and the notion of quantity in physics, which goes back even earlier, to Descartes. They are so closely related that we must discuss them together and at the same time alternately: We begin in this chapter with the notion of quantity and the first results we obtain bring a significant conclusion on nonlocality. The next chapter will pursue the discussion of quantity. This splitting of issues has the advantage of cleanly separating two questions, which look very different from the standpoint of philosophy, whereas they are deeply related from the standpoint of quantum mechanics.

## DESCARTES ON QUANTITY

The association of physical reality with numerical quantities is certainly not innate in mankind. It does not come directly from the mechanisms of our perception and it is definitely a philosophical conception, now become cultural. If one leaves aside Pythagorean mysticism and the usual premonitions preceding the formulation of a great idea, this one was first clearly formulated in Descartes' *Discourse on the Method*. This is very late for a belief that has nearly become universal by now, and one must clearly associate its success with the expansion of science. Astronomy had earlier given a suggestion of the importance of translating an observation into numerical records, but the rise of experimental science around 1600 brought it down from the heavens to earth.

Galileo had shown that science results from careful measurements, implying that physical laws are expressed in the *lingua mathematica*, but he had not carried this idea down to foundations. After Descartes, it became common knowledge that every measurement (as it could be done at that time) ended in a position measurement: the position of an index on a ruler or a dial (now, it is often directly registered as a number in a computer memory). Since Euclid and earlier, position had provided a basis for the concept of a real number, and this is how a direct conceptual link between physical reality and mathematics came out explicitly. Descartes' triumph was the invention of analytic geometry, which occurs in the *Discourse*, showing that everything in geometry reduces to numbers. It may be noticed that his reflections extend also to what we would call the mechanisms of life, and Edmund Husserl was certainly right in calling the "Cartesian program" the philosophical proposition according to which everything in reality is completely controlled by quantitative laws.[1] Most of us learned of this idea in a pervasive way from teachers who did not really teach it, but transmitted it as an unformulated and obvious belief, in the same way as they received it.

"Quantitative laws" suppose the notion of quantity, identified with number; this is much more restrictive than the idea of "mathematical laws," if one thinks of mathematics as a pure science of relations and not as the science of number and geometrical objects that it was long ago. This difference is

---

[1] See Husserl (1970). Husserl wrote a basic paper on the mathematization of nature in 1928, which has been included by editors in the 1970 book collecting his ideas on science and phenomenology.

completely ignored in Heidegger's critique of the Cartesian program, as far as I know, and one cannot use it therefore as a convenient philosophical reference. The deepest forms of questioning originate anyway from quantum physics and they will be among our topics in this chapter and the next one.

## REALITY AND PHYSICAL QUANTITIES

Quantum physics led its founders, particularly Paul Dirac, to a deep reflection on quantity and the corresponding physical concepts. Everybody agrees that everything in physics is sanctioned by measurements and measurements yield quantities, i.e., numbers. But a measurement is performed by means of some apparatus, which is necessarily macroscopic and therefore made of a huge number of atoms, assembled in a complex system. Any system of that kind (and almost any macroscopic one) has very specific properties, among which the main one is to exhibit the classical behavior from which stems our innate notion of reality. The origin of that property has only recently been understood and is due to a "decoherence" effect, which will be discussed later, but the underlying quantum concepts, the primary and efficient ones, cannot be reduced to sheer numbers.

This impossibility will be explained in this chapter and its mathematical significance in the next one, the main question being: Can one speak of physical quantities as if they were simply numbers? A question of vocabulary then arises immediately: are the words "quantity" or "variables" sufficiently different from the word "number" to allow a clean discussion? A virgin word would certainly be better and Dirac accordingly proposed "observable," whereas he had used first "$q$-number" (as opposed to an ordinary number, which he called a "$c$-number," "$q$" and "$c$" referring of course to "quantum" and "classical"). The first proposal was fine in the framework of physics, but it could not be conveniently exported to mathematics.

The second proposal was more attractive, although it also has its drawbacks. The suffix "able" is nice, since it suggests that the thing "can" be observed, which does not mean that it "is" observed. But can it always be observed, even as a matter of principle? One would have to say what principle decides it, and von Neumann thought he had an answer. He introduced formal measurement devices, obeying quantum laws although highly unrealistic, and he showed that such a device can certainly measure anything to which one attributes the quality of a physical concept, for instance, position $x$, momentum $p$, on a function involving both of them under some well-defined conditions on the symmetry of products. It looked fine, but in the last pages of his *Mathematical Foundations of Quantum Mechanics*, where that was explained, he fell upon an apparently sweeping paradox, which is best known under the name of Schrödinger's cat dilemma. As a matter of fact, not every observable can actually be observed: the extent of concepts is wider than the experimental possibilities!

The concept of an observable according to von Neumann is necessary, in any case, for the sake of mathematical consistency and it will be explained in the next chapter. It has a strong link with numbers although it is not itself a number, as shown by the spectral theory of operators in Hilbert spaces (I do not intend to go as far in the mathematics of quantum theory as to develop that). Since we are presently discussing a matter of vocabulary, it seems that the neologism "numberable" would be the most convenient word for expressing the sense of the quantum concept most akin to a physical quantity. We will hold on, however, to "observable" after the present note of caution, since it is now used universally.

## Can One Measure Two Quantities Simultaneously?

We saw an example of a position measurement chessboard game, in which a net was used. As a matter of fact, the net was

introduced only to make that kind of process visible and to materialize the squares of uncertainty in the measurement. The
same thing can be done for a velocity measurement, although I
will not try to make it palpable. Rather than the velocity $v$ of the
particle, moreover, it will be convenient to envision a measurement of its momentum, $p = mv$, where $m$ is the mass (the momentum being in that case a two-dimensional vector).

Rather than picking out the particle histories whose
"clone" is in a definite square at the time $T$ when the measurement is performed, we may split the two-dimensional plane in
which $p$ is defined into "momentum squares," indicating the
possible results of a measurement together with their uncertainty. We then pick up the clones having their momentum in
such a square at time $T$, as we did for a chessboard (position)
square. All the previous steps, the calculation of the action, the
phase of a Feynman history, and the summation of the clone
pointers to give a probability amplitude for every possible measurement result, will be essentially the same. One thus obtains a
probability distribution for the results.

Well, that's fine, you might say. Let us go one step further and get a probability distribution for a simultaneous
measurement of position and momentum. We split the position plane (the chessboard) into squares, we do the same thing
in the momentum plane to split it into momentum squares,
and we define a distribution probability for this new measurement. There is no special difficulty. That can be done and we
may as well lighten the notation by denoting a chessboard
square by the position $X$ of its center (notice that this $X$ is not
continuous and its values form a lattice). We do the same for a
momentum square, labeling it by the coordinates $P$ of its center. We thus obtain a probability distribution $Q(X, P)$ for position and momentum, just as we previously got a probability
distribution $R(X)$ for the position squares.

Simple, isn't it? But let us check. Surely the $X$-probability
distribution $R(X)$ should result from the $(X, P)$-distribution
$Q(X, P)$ by summing it over all the squares $P$. This kind of

summation is supposed fundamental in probability theory, and one says that the measurements are independent. To give an example, let us suppose we pick up a ball at random from a box, some balls being red and others green, the surface of each ball being either rough or smooth. We know how many balls are red and rough, green and smooth, and so on. Then what is the probability of the ball we draw being red? The obvious answer is that it is the sum of the probabilities of the ball being red and either rough or smooth; this answer does not assume that the color and the roughness of the balls are uncorrelated: there could be, for instance, a larger proportion of rough balls among the red ones than among the green ones. Let us assume then that we don't look at the ball nor touch it and we ask two people to "measure" it. The first person is color blind and touches the ball: she finds whether it is rough or smooth. The second person looks at the ball from a distance and sees its color.

We expect to get the same probability for the red color when the two persons respectively touch and see the balls, or when only the second one looks at it and the first one is absent. It looks obvious, because, after all, only what the second person says is significant. When asked why we think so, we would answer that touching a ball does not change its color! Touching and seeing are two independent operations, made by two independent and compatible "instruments," namely, the hand of the first person and the eye of the second one; this again does not assume that roughness and color are uncorrelated. In the present case, $X$ stands for color and $P$ for roughness, and we are assuming that they can be *measured* independently, this independence in the measurements implying, in analogy with the balls, that $R(X)$ is the sum of $Q(X, P)$ over all the values of $P$. But that is not what the calculation gives. This means that we were wrong when assuming that $X$ and $P$ can be measured independently. If we let the squares on the chessboard and in momentum space become infinitely small, it means that position and momentum cannot be measured independently at the same time.

## *The Uncertainty Relations*

The discussion we just gave was based on the fundamental rules of quantum mechanics, in the form given by Feynman histories, although we did not show the explicit calculation. Other expressions of the principles of quantum mechanics are equivalent and they yield the same conclusion of course. There was much fuss in the physicist community when Heisenberg discovered it. Bohr and Heisenberg tried to convince themselves by imagining various sorts of apparatus which would have a chance of measuring position and momentum simultaneously. Heisenberg proposed a tremendously powerful microscope showing the position of a fixed electron, but the photons of the light falling on the electron changed its momentum, while diffraction implied that the position was inevitably blurred. With light of a higher frequency, the position was better known, but the photons had a higher momentum, and so the uncertainty in the electron momentum increased.

Bohr considered a clever device, which did not rely on the wave-particle—or field and photon—character of light. It consisted of a screen with a hole allowing an electron to get through. One could thus know what the transverse position of the electron was when it crossed the hole. The screen was suspended on springs. The electron momentum was initially directed toward the screen and one could know the change in its momentum when it crossed the hole: it gave the screen a transverse momentum, opposite to its own. A look at the oscillation of the screen could tell us what this momentum was. It looked therefore as if one measured simultaneously the transverse position and the momentum of the electron. But the uncertainty relations are universal and they also apply to the screen. The smaller its own initial momentum, the better one knows the electron momentum, but the uncertainty relations imply that the uncertainty in the hole position is larger, and so the knowledge of the electron position is also less. Finally, one could not beat the uncertainty relations for the electron; they were perfectly consistent. Einstein tried to devise counterexamples, some of them very clever, but they only confirmed the conclusion.

The uncertainty relations do not belong to the fundamental principles of quantum mechanics. They are consequences of these principles and they have a precise expression, which we need not write. In the case of our chessboard game, they mean that the sizes of the two kinds of squares, in position and momentum space, cannot be made arbitrarily small together. The product of their sides must remain larger than Planck's constant $h$, and one cannot do better than that in the simultaneous knowledge of position and momentum. The concepts of position and momentum cannot therefore reduce to the concept of numbers.

## ON LOCALITY

Every object, according to the classical conception of reality, is in a well-defined place at any given time. Kant said that our reason wants it. Our eye sees it, obviously, but it cannot be true in a quantum world. This unwieldy assertion is practically an axiom in the conception of Feynman histories, but it is probably better understood as a consequence of the uncertainty relations. Suppose that it were true that a particle really has a position $x(t)$ at any time $t$. We could then derive a definite velocity $v(t)$ from a change in position during a short time, and thereby a definite momentum $p(t)$. That would clearly be at variance with the uncertainty relations. This means that we cannot even *think* of a particle as having a definite trajectory and, as a matter of fact, we must think of the huge amount of Feynman histories as equally significant events (which does not mean equally probable events, but let us leave that aside).

This assertion should be taken, however, with a grain of salt and not applied blindly. It means that we cannot think of a sharp position $x(t)$; it does not mean that we cannot envision a fuzzy position. There are many instances when one can speak of the position and the trajectory of a particle, usefully and legitimately. One can even see a trajectory in a bubble chamber or a wire chamber, two kinds of detecting instruments that were and are extremely important in particle physics. When a high-velocity charged particle crosses a bubble chamber, for instance, it ionizes some atoms in its neighborhood, i.e., pulls away electrons from them. The chamber is filled with a liquid, in thermal conditions such that gas bubbles are produced by the charge of these free electrons, and the bubbles soon become big enough to be seen on a photograph, indicating through their alignment a trajectory of the particle. This trajectory is not sharply defined, however, if only because the bubbles are macroscopically big: it is a fuzzy trajectory.

We might also comment why it makes sense to speak of a molecule having a reasonably well-defined position in a crystal, or an atom inside a molecule (at least in the usual cases). An atomic electron, however, is anywhere, or one might say everywhere, in the vicinity of the nucleus. The difference is a matter of precision in the position one is talking about and the physical condition of the particle one is considering. Every case must be dealt with specifically.

It should also be mentioned for completeness that, although we cannot tell a priori where an atomic electron is, we could measure its position, arbitrarily sharply in principle. We can send, for instance, a high-energy photon onto the atom and we can say "here it was" after the experiment. But this measurement is a random process and when it is repeated many times under identical conditions, the outcome is only a probability for the electron to be found in such or such a place, exactly as when we were throwing a net on a chessboard.

## On Identical Particles

We mentioned in chapter 2 that our consciousness would not put up with the presence of ambiguities in what we see: reality must be sharp, or so it is classically. Leibniz made a careful philosophical analysis of the consequences of that presumption and he stated that any object is always distinguishable from another object, no confusion being conceivable. This idea was central in his philosophy of knowledge and his metaphysics. Two principles, continuity and distinguishing, were introduced in his *New Essays on Human Understanding* as the foundation of philosophy. This question is not therefore a side issue in the history of the idea of reality, and it will be interesting to compare Leibniz's analysis with what quantum mechanics says on that topic.

Let us first state that the impossibility of distinguishing two particles of the same species—two electrons for instance—is one of the basic principles of quantum mechanics. It is therefore exactly opposite to Leibniz's assertion. It is expressed mathematically by the permutation symmetries of the wave functions, which we do not write down. Their physical consequences are anyway convincing enough to make clear why they hold, the principle of indistinguishable particles explaining a multitude of facts indeed, among which a few may be mentioned: Why there are electron shells in atoms (from which the chemical properties follow); why a table is hard, although the atoms in it consist mainly of vacuum; why some substances conduct electricity and others are insulating; why there are semiconductors (and so transistors and computers); why lasers work; why nuclei have a size; why some stars become white dwarfs and some others neutron stars. As a matter of fact, the list could involve almost as many items as an encyclopedia of physics!

Leibniz asserted that two objects, even when they are perfectly identical, can always be distinguished because each one of them has a well-defined position in space at some given time. One can then distinguish them at that initial time through their position and give them a distinctive mark, calling for instance one of them Alice and the other Bob, or give them labels A and B. Their motion can never bring them to exactly the same place and, since this motion is continuous, one is able to follow them individually and never confuse them. They are therefore distinguished forever by virtue of two properties of classical reality: continuity and locality.

There is a similar analysis in Dirac's *Principles of Quantum Mechanics*. He agrees with Leibniz on the possibility of an initial distinction of two electrons. One can in principle measure their positions at an intial time and give them a distinctive label. But one will lose track of that label later, because the two particles have no well-defined trajectories. It is therefore possible to escape Leibniz's argument and quantum particles *may* be undistinguishable. Many experiments and physical consequences show that they are (that was certainly bad news in the best possible world, which was supposed to be ours according to Leibniz).

# More about Physical Quantities

The aim of the present chapter is simply to complete the discussion of observables, which were introduced in the previous one. The main property underlying their peculiarities will be found to be their lack of commutation. It turns out, however, that the framework of Feynman histories is not especially convenient for discussing observables and their multiplication properties. This drawback can be avoided in other approaches to quantum mechanics which are introduced here, particularly the Hilbert space framework in which observables and their noncommutation are easy to grasp.

Although the question of noncommutation is certainly important for the philosophy of physics, it is not essential for the purpose of physism within the philosophy of mathematics. We shall, however, need some knowledge of other approaches to quantum mechanics in the next chapter, and they will be conveniently introduced in the present one.

### The First Occurrence of Observables

When Heisenberg discovered his formulation of quantum mechanics, he considered right from the beginning that the task had to do with the concept of physical quantities. He started from an idea—which he attributed wrongly to

Einstein—according to which every mathematical notion entering in a theory should express something that can be measured experimentally. We shall not pause to discuss this assertion as an epistemic statement, the only important point presently being that, right or wrong, Heisenberg got an inspiration from it. He found with some surprise that the physical quantities he had introduced for the position $X$ of an electron and its momentum $P$ did not commute or, more precisely, the product $XP$ was not equal to the commuted product $PX$. He was not deterred, however, by this strange result, because it implied remarkable experimental consequences, which had never been explained previously.

Inventing a new physical theory—or a new mathematical theory—together with the necessary new concepts is one of the greatest adventures of the human mind. It does not look very different in theoretical physics or mathematics and I thought that at least one example should be given in this book. Since I would not dare to do it for something mathematical, and since moreover it must be nontrivial to be significant, I propose to consider the first *nonclassical* theory, Heisenberg's, as a paradigm.[1] It has unavoidably some technical aspects, but I hope they will not deter readers, or they can of course skip it.

### Heisenberg's Matrices

Heisenberg could take for granted the existence of stationary states of the atom with definite energies, which had been introduced by Bohr and confirmed experimentally. Bohr had also shown that, when a photon is emitted, the atom "jumps" from a state $n$ with energy $E_n$ to a state $k$ with a lower energy $E_k$, and the frequency of the emitted light is

$$v_{nk} = (E_n - E_k)/h. \tag{6.1}$$

[1] Louis de Broglie's idea of the wave function came before Heisenberg's work, but it appeared first as a remarkable physical possibility, not as a reconsideration of the foundations of classical reality.

Heisenberg knew also that the emission of light is controlled by the electric dipole of the source, according to the Maxwell-Hertz theory of light. The electric dipole $d$ of an atom is simply related to the position $x$ of the electron relative to the nucleus by $d = ex$, $e$ being the electron charge (we consider a hydrogen atom with a unique electron for simplicity). So the position of the electron controls light emission and, according to Heisenberg's epistemic preconception, emission is the only physical observation allowing one to get at the electron position. But we don't get a number $x$, we get only a collection of pieces of information concerning the various transitions $n \to k$. What could that information be?

Heisenberg considered that it must consist of a collection of numbers $\{X_{nk}\}$. One must mention that he had previously worked out a theory of the index of refraction of a substance, in collaboration with Henrik Kramers, and they had encountered a collection of numbers $\{D_{nk}\}$ involving the dipole moment. Their calculation was done in the framework of the Bohr-Sommerfeld theory, which supposed that the electron always had a definite position, and these numbers expressed its motion when it jumps from a stationary trajectory $n$ to another $k$ during emission. Heisenberg rejected this "old quantum theory," which had meanwhile been criticized by Bohr himself as inconsistent, but the step of considering such a collection looked convenient.

But what could replace the fact that the earlier results depended on the electron motion, or more abstractly on the time dependence of $x(t)$? The relation (6.1) gave a hint. The frequency of light enters in the time variation of a monochromatic light wave as a factor $e^{-2\pi i v t}$, which suggested a time dependence of $X_{nk}$:

$$X_{nk}(t) = X_{nk} \exp[-i(E_n - E_k)t/\hbar]. \tag{6.2}$$

Heisenberg invoked hand-waving arguments involving resonance effects to justify this assumption, but what else could it be? There was something puzzling in the fact that one was dealing with complex exponentials and not with sines and cosines, but that worry came only later.

The final test of a physical theory is experiment, and Heisenberg had to draw predictions from his construction. As a matter of fact, he had embarked on it with the prospect of computing the intensity of spectral lines, which was yet unexplained. The Maxwell-Hertz theory said that the emission intensity of a classical source of light is proportional to the square of the dipole moment, or equivalently the square $x^2$ of the position. What square? It was surely something physically meaningful during emission, and that implied the introduction of a new collection $\{(X^2)_{nk}\}$. What could be the relation between $(X^2)_{nk}$ and the quantities $X_{jl}$? Dimensional considerations implied immediately that $(X^2)_{nk}$ had to be a quadratic function of the $X_{jl}$, and the relation (6.2) implied restrictions on it. The time dependence should remain consistent and the answer had to be therefore

$$(X^2)_{nk} = \sum_j C_{nkj} X_{nj} X_{jk},$$

involving some unknown coefficients $C_{nkj}$. What could they be? Heisenberg invoked Bohr's correspondence principle, which was again some sort of hand waving. He might have said also that algebra cannot depend upon the atom to which it is applied and the coefficients had to be universal. Anyway, that led to the guess

$$(X^2)_{nk} = \sum_j X_{nj} X_{jk}. \tag{6.3}$$

Then what about dynamics? The velocity of the electron was also certainly replaced by a collection $\{V_{nk}(t)\}$ involving the time derivatives of $X_{nk}(t)$, i.e. (considering the momentum $P$ rather than velocity),

$$P_{nk} = -[im(E_n - E_k)/\hbar] X_{nk}. \tag{6.4}$$

A serious difficulty arises at that stage: when multiplying position and momentum, one must decide the place of $X$ and $P$ in the expression generalizing equation (3). The product $XP$ is not equal to the product *PX*. Youth is bold. Heisenberg was twenty-three years old and he did not waver at that junction; he proceeded anyway, through guesswork and calculations, which yielded predictions for emission intensity. And these predictions agreed with data!

Max Born, his adviser, was deeply impressed. He told Heisenberg that some mathematicians had previously introduced similar products, and they called the corresponding objects "matrices." Heisenberg's matrices were infinite, but who would mind? They joined forces with Pascual Jordan and the trio rapidly made tremendous progress. They found that, if one assumed only a general commutation relation between $X$ and $P$, given by

$$[X, P] \equiv XP - PX = i\hbar, \tag{6.5}$$

one obtained a wonderful dynamical theory, with an impressive amount of predictions, all of them consistent with a lot of already known experimental results and many unexplained ones.

We may stop the story there. It shows a magnificent example of the mixing of intuition, boldness, temporary blissful carelessness, wrong though creative prejudice, in one word illogical invention, which may sometimes produce a new theory in physics or mathematics. Logic comes afterward, with the search for consistency or for a contact with already existing foundations. Anyway, it gave an answer to the problem we started from: physical quantities (observables) are not numbers, and they do not commute.

## WHY NONCOMMUTATION?

Sam Schweber mentions in his book (1994) an unpublished lecture by Julian Schwinger, which was given in the early 1960s on the subject of quantum physics and philosophy.

Schwinger explained that numbers come out in microscopic physics when, and only when, a measurement is performed. He made a point that a measurement "strongly perturbs" the system that is measured, because of its smallness.[2] He then explained noncommutation:

> If we once recognize that the act of measurement introduces in the [microscopic] object of measurement changes which are not arbitrarily small, and which cannot be precisely controlled . . . then every time we make a measurement, we introduce a new physical situation and we can no longer be sure that the new physical situation corresponds to the same physical properties which we had obtained by an earlier measurement. In other words, if you measure two physical properties in one order, and then the other, which classically would absolutely make no difference, these in the microscopic realm are simply two different experiments.
>
> So therefore the mathematical scheme can certainly not be the assignment, the association, or the representation of physical properties by numbers because numbers do not have this property of depending upon the order in which the measurements are carried out. . . . We must instead look for a new mathematical scheme in which the order of performance of physical operations is represented by an order of performance of mathematical operations. (Schweber 1994, 361)

Schwinger (at least in that lecture) followed Bohr's interpretation of quantum mechanics, which gives a fundamental role to measurements. It is an experimental fact that two different measurements may not commute, and we might infer from that that the two measured observables do

---

[2] As a matter of fact, one can sometimes perform nonperturbing (so-called nondestructive) measurements, but these techniques are rather recent and were not yet used when Schwinger gave his lecture.

not commute. But noncommutation was discovered before any experiment of that kind was made and it was therefore *predicted* that two measurements generally do not commute. Atoms were there long before any human being made quantum measurements and, surely, reality is independent of our intention of looking at it. Measurements are the bread and life of physicists, but nothing more.

I won't try to tell therefore "why" observables do not commute. I don't know, and I don't know anybody who does (which is a much more valuable statement). One can relate that property to other features of the quantum laws and thereby ascertain the depth of their consistency, but that cannot be an explanation of any little piece of them. We cannot explain why the laws are what they are. God only knows, but we don't, and perhaps, like the rose of the poet, He has no "why."

## The Case of Wave Mechanics

Two years earlier than Heisenberg's work, Louis de Broglie had suggested a very different approach. He was inspired by experimental data on X-rays. He knew that these rays show a clear wave character when they are diffracted by a crystal, but they also leave straight-line tracks when they cross a stack of photographic plates. Could it be that this duality of wave and particle features is universal, and was there also a wave acting behind the electron? Using the theory of relativity and an analogy with Einstein's formula for the energy of a photon in terms of the light frequency, de Broglie proposed a simple relation between the wavelength of the supposed wave and the momentum of the particle in the case of free motion.

Schrödinger elaborated de Broglie's proposal a few years later. It was one year after Heisenberg's work, and Schrödinger expected hopefully that a sensible form of quantum physics would emerge from the wave approach and would get rid of the incredible noncommuting matrices. The first steps in his investigation did not encourage this optimism. The wave function generally had to be complex. In the case of an atom, it could not be a simple wave existing in three-dimensional space (as de Broglie had assumed), but it had to depend on the positions of all the atomic electrons. These unexpected features pointed toward a purely mathematical object rather than a classically realistic concept, but it took some time before that consequence was fully realized.

Meanwhile, Schrödinger had proposed a principle for expressing dynamics as a time evolution of the wave function. As a matter of fact, it was more an equation than a principle that could be expressed in words; it was the famous Schrödinger equation, which is still used after three-quarters of a century with an ever-increasing scope. Schrödinger's invention of this equation (which I will not describe here) relied on an empirical recipe: He used the usual formulas for the dynamics of point particles interacting through Coulomb potentials, and he replaced a momentum $P$ everywhere by a derivation operator $-i\hbar/\partial x$. Why did he? Like Heisenberg previously, he had only clever though hand-waving arguments (one can never tell why a law is what it is!).

He had apparently got rid of matrices, but not of noncommutation. The operator $X$ consisting in multiplying the wave function by its position variable $x$, and the differential operator $P$ replacing the momentum again satisfied the prophetic equation (6.5). They did not commute, and the most puzzling property of the Heisenberg matrices was found to be inescapable. Dirac and Schrödinger himself proved soon after that the two formulations of quantum mechanics, in terms of either matrices or wave functions, are actually equivalent from a mathematical standpoint: matrices act on vectors, and the vectors are wave functions.

## The Hilbert Space Formalism

The idea of considering some sets of functions as a space yielded one of the most powerful tools of functional analysis in the twentieth century. Paul Dirac applied it first to quantum mechanics, in a somewhat intuitive fashion, and John Von Neumann did it with the full arsenal of rigorous mathematics. Wave functions are square-integrable, since the integral of their square is the total probability (namely, 1) for particles to be anywhere, and the set of square-integrable functions (using the Lebesgue integral) is a Hilbert space. A Hilbert space is defined as a complete complex vector space that is endowed with a scalar product, but the present book is not the place for elaborating on this definition or its many mathematical consequences.

A hint of this framework may be obtained, however, by going back to our pet chessboard example. We consider the case of a board with a finite number $n$ of squares. We saw how one can define a set of $n$ amplitudes on the squares; let us call them $y_k$ $(1 \leq k \leq n)$. We will consider these complex numbers as the coordinates of a vector in an $n$-dimensional vector space. This is in fact an $n$-dimensional Hilbert space, which is most simply defined as an $n$-dimensional Euclidean space in which the coordinates have been made complex. The Euclidean scalar product is extended to complex vectors in the following manner: Let $Y$ be a vector with coordinates $y_k$ and $Z$ another vector with coordinates $z_k$; their scalar product is the number

$$(Y, Z) = \sum_k y_k^* z_k.$$

One then defines an operator as a linear transformation on the vectors. It can always be written as a square $n \times n$ matrix acting on the vector coordinates. If we denote the action of such an operator $A$ on a vector $Y$ by $AY$, the adjoint $A^*$ of the operator $A$ is defined by the condition $(A^*Y, Z) = (Y, AZ)$, the matrix of $A^*$ being obtained by transposing the matrix of $A$ (i.e., taking its symmetric with respect to the main diagonal) and taking the complex conjugate of the matrix elements. One therefore defines a self-adjoint matrix by $A^* = A$, and the Von Neumann definition of an arbitrary observable boils down in our example to an arbitrary self-adjoint matrix. The product of two observables is then in general, once again, noncommutative.

We said that an observable is not a number, but it is associated with numbers: the values it can exhibit under a measurement. This relation between an observable and numbers is explained as follows: In an $n$-dimensional Euclidean vector space, a symmetric (real) matrix $A$ is known to have $n$ orthogonal eigenvectors $Y_k$ and corresponding eigenvalues $a_k$, the real numbers such that $AY_k = a_k Y_k$. The same is true in a complex Euclidean space in the case of a self-adjoint matrix $A$, the eigenvalues being real numbers and the orthogonality of two eigenvectors amounting to the vanishing of their scalar product. The eigenvalues are the values of the observable $A$ that can be found in a measurement, and their set is called the spectrum of $A$. Similar results hold in a general Hilbert space, with, of course, a few niceties, but to say more would bring us into a quagmire of topology and I hope that that will be enough, if not already too much. Fortunately, although all this is very important for doing physics, we shall be able to skip it during our excursion: the philosophy of quantum mechanics would not allow that omission, but the philosophy of mathematics can make it.

## CHAPTER SEVEN

# On the Extent of the "Lingua Mathematica"

This chapter provides an important step in the construction of physicism. It deals with the well-known idea of considering mathematics as the language of nature: the *lingua mathematica* in which the book of nature is written, according to Galileo. Although the chapter begins in a light-hearted mood, we shall soon find ourselves brought to questions that are seldom raised. I hope their discussion will convince the reader that no sound law of nature would exist if mathematics were inconsistent. Moreover, the whole corpus of mathematics (or at least the most part of it) is involved in this coexistence of physics and mathematics.

## MATHEMATICS AND THE LANGUAGE OF PHYSICS

Since Galileo, mathematics has been recognized as the language of nature, and particularly physics. Let us then ignore our sense of beauty, of intellectual enjoyment when doing mathematics, and ask only the question: What would it mean if we considered mathematics *uniquely* as the language of physics? To make this strange idea palpable, let us imagine another mankind on another planet, intensely interested in nature though with a strictly empirical mind. The people in it can enjoy nature, but nothing else, no art for instance; they contemplate nature and they have discovered that precise laws rule it, which they study thoroughly. Maybe they are also interested in the practical applications of science, like many people on earth. Anyway, they experiment and they learn.

They are not unable to do math, far from that, and they have soon invented it for their own purposes. They consider it, however, as an inconvenient necessity, like the necessity of cleaning a test-tube after an experiment. They know how clean a good test-tube should be and they resign themselves to digging further into mathematics, not for the fun of it but as some unavoidable asceticism. Everybody on that planet has learned the great experiments in some university, with an incommensurable pleasure, but most people are reluctant to learn some math, except for a few ascetics who are highly paid for that, even more than garbage collectors—since garbage collectors enjoy a closer contact with nature.

Now suppose that a small group of young ascetics, calling themselves Bourbakos, join together and say: "We must get to the bottom of this relation between the laws of nature and mathematics (the hell with it!). We must show the great consistency of the knowledge we have reached by writing a book—let's call it *Elements of Science*—or if necessary a long series of books. It will start from the alphabet of the language of science, by which we mean the primary concepts, either mathematical or physical as needs may be, consistently, which means that nothing will be introduced until the necessary previous notions have been cleared up. No mention will be made of experiments (showing by the way that our ascetics are not real physicists); concepts and consistency must rule!"

We may suppose for convenience that the state of knowledge in physics among the people of that planet is essentially similar to ours at present. They begin with mathematical logic and set theory and they have a long discussion about the proposal of one of them, by the name of Zermeli, who proposes what he calls an axiom of choice. They decide not to decide and let time do the work for them. They introduce discrete topology, integers, arithmetic, and, because they are extremely conscientious, they write many exercise books to investigate the extent and the boundaries of what has been introduced, to make sure that nothing potentially useful for explaining the laws of nature has been forgotten. Their exercise books are certainly as good as the best treatises of pure mathematics on our earth, although written with much suffering.

Nothing about physics is written for a very long time. It would be tempting to say something about classical physics, but that is impossible. It stands ultimately on quantum principles and one must wait for that. There was much enjoyment when they could at last formulate differential geometry. It was at last possible to mingle it with the theory of space-time! They arrived sometime later at Hilbert spaces and it was finally possible to speak of quantum mechanics. Particularly hard exertions in asceticism allowed them to obtain an axiomatic foundation for what we call Feynman histories (no human has done that yet). Then, they could develop theoretical physics along two complementary directions. One direction was particle physics, which reached approximately the level of our standard model of quarks and leptons, with interludes for developing necessary distribution theory, higher group theory, fiber spaces, and a few other jewels created from suffering. The other direction was toward classical macroscopic physics, including microlocal analysis, decoherence (they have a good theory for that), and probability calculus.

Bourbakos are very thorough. They know that a mathematical theory will be inconsistent if any part of it using the same axioms fails. This is why they are so careful in pursuing endlessly useless problems, such as the one we call Fermat's last theorem, for instance: the poor people are unable to see any beauty in it, but they encounter new concepts during their search, and nobody can tell whether these concepts have any relation with the laws of nature.

The conclusion of this little apologue is clear: If one reduced mathematics to the unique and modest role of a language of the laws of physics, the present state of theoretical physics and simple consistency requirements would not appreciably modify the present corpus of mathematics.

## The Extent of the "Lingua Mathematica"

We may now come back to Earth and try to give a more critical account of the last statement. It can be conveniently formulated in five points: (i) The formulation of physical laws—as we know them presently—definitely requires mathematical concepts. The closeness of physical and mathematical procedures is even stronger when one derives the consequences of the laws, since such a derivation must usually follow the pattern of a mathematical proof. The consequences that are thus obtained from the basic laws (also called "principles"[1]) constitute the predictions of a theory, and they can be directly compared with observation and experiment to check the veracity of the theory or use it for applications. (ii) A physical theory must be logically and mathematically sound, i.e., it is of no value if it is inconsistent in a mathematical sense. (iii) When a physical theory (or a part of it) requires mathematics in the formalized corpus (i.e., belonging to the wide part of mathematics with an axiomatic foundation), one can make the axioms necessary for the theory explicit, at least in principle, and follow the unfolding of ideas from these axioms into the mathematical corpus. The physical laws, as we know them presently, reach through this process into an extremely large part of the mathematical corpus, and maybe all of it. (iv) If an inconsistency were to occur in such a part of mathematics, the corresponding laws would become nonsense. (v) The consistency of mathematics is therefore tantamount to the existence of mathematically expressible laws of nature.

---

[1] The distinction between principles, derived laws, and empirical rules is discussed in chapter 14 of my book *Quantum Philosophy*.

Some points in this list of ideas would be worth more discussion. For instance, what about the nonformalized parts of mathematics, which have been and still are used in physics? Feynman histories, for instance, represent probably the most important mathematical notion in physics that does not yet have a complete axiomatic foundation.[2] But we may also turn to the past. Heaviside calculus was used with success in the analysis of electric circuits long before the discovery of distribution theory, which gave it a sound mathematical basis. The same is true of Dirac's famous delta function. The expert practitioners in theoretical physics developed on these occasions a knack for avoiding pitfalls, and their success was already an indication that a sound mathematical theory stood behind. Many concepts in quantum field theory are now formulated in terms of Feynman histories, and most calculations use them. The success of particle physics with gauge theories and the standard model of quarks and leptons is so tremendous that the bet for Feynman histories to hold a future grand mathematical theory in reserve has much higher odds than its incompatibility with the rest of math.

Another interesting question is raised by point (iii): what exactly is the extent of the present mathematical corpus that is in relation with the mathematics of physics? I cannot say I have analyzed this question carefully, but I considered it from time to time when reading papers in theoretical physics or mathematics. I am sure that no theoretician, no mathematical physicist or mathematician well aware of physics, will contest that most of mathematics is affiliated in that way to physics. Two parts only, to my knowledge, might raise discussion: (1) Does physics need the axiom of choice? (2) Does it

[2] Pierre Cartier and Cécile Morrette–De Witt have gone a rather long way toward a mathematical understanding of Feynman histories, but not yet far enough to obtain a complete axiomatic foundation.

need some highbrow mathematics, related to deep logical considerations?

When speaking of logically inspired highbrow mathematics, I am thinking of a few notions that could be of interest for an average physicist, and not of higher logic itself. As a matter of fact, I prefer to confess that I am a wishful admirer of higher logic, but a very ignorant and stupid one, and I am rather sorry that I could not avoid that question. The examples I am thinking of are particularly nonstandard analysis, Chaitin's Omega numbers, and Cohen's alternative set theories with or without the axiom of choice or the continuum hypothesis. Have they anything to do with physics or at least the language of physics? I cannot see a serious hint of that. Is such a relation impossible forever? See how crafty I became after reading philosophy books: the answer is obviously "No," and so I can go on. I will not have to comment, at least at present, upon a part of mathematics that is apparently foreign to nature and a part associated with it. But let us rather go to what is immediately worth comment.

## About the Axiom of Choice

The question of the relevance of the axiom of choice is easier: there is a place in physics where this axiom clearly cannot be left out. It comes from a comparison of the basic axioms of quantum mechanics, as introduced, for instance, in Dirac's book:

1. The superposition principle (holding for wave functions or Hilbert space vectors and resulting in interferences)
2. Noncommutable observables
3. Symmetries (of either wave functions or operators) under an exchange of identical particles
4. Dynamics: It can be equivalently stated on wave functions

(Schrödinger equation) or operators (as first done by Born, Heisenberg, and Jordan)

5. Probabilities

Nothing will be said here of measurements, since the corresponding rules are now known to derive from these five principles. Principles 3 to 5 rely on Principles 1 and 2, and each of them has two versions, applying, respectively, to wave functions or observables (this is not quite obvious for Principle 5 and probabilities, but never mind).

One may ask therefore the following question: Which is the more basic principle between 1 and 2? In the Schrödinger approach and the Hilbert space formalism, wave functions or vectors come first, and then observables can be introduced as operators acting on them. Could it be that one might also start from an algebra of observables and build a Hilbert space from it?

### Two Abstract Mathematical Constructions*

Von Neumann asked that question and went a rather long way toward answering it.[3] He introduced for that purpose the concept of $C^*$-algebra, which we do not need to define mathematically in a precise way. Let us only say that the $q$-numbers composing it (this old expression by Dirac is convenient for such abstract purposes) can be added, multiplied together, multiplied by complex numbers, and a positive quantity (a norm) yields a topology on the algebra.

Two different approaches have been used for introducing Hilbert spaces in which $q$-numbers are represented by operators. The first one, due to Von Neumann, inverts the transition from operators to observables: some elements of the algebra are considered as observables and a clever construction defines a Hilbert space in which they act as operators.

[3] A standard reference on $C^*$-algebras remains Dixmier (1964).

The second method selects a maximal commutative subalgebra and concentrates on it. The study of commutative C*-algebras has been developed (for mathematical purposes) by Izrail Gelfand and, independently, by Irving Segal. Their main result was that a definite geometrical space is associated with the algebra, the $q$-numbers becoming simply functions on that space.[4] One can then construct a Hilbert space using this "configuration space" as the domain on which wave functions are defined.

### A Conclusion about Physics and the Axiom of Choice

We just discussed briefly two methods for constructing wave functions (or more exactly Hilbert space vectors) when one is given a family of observables. The two methods, one by Von Neumann and one by Gelfand, have different applications in mathematics and both involve highly technical developments. A simple and significant remark emerges, however, from a consideration of the proofs. In Von Neumann's theory, one constructs the vectors spanning a Hilbert space by means of the Hahn-Banach theorem (which deals with a positive convex functional), and nothing can be obtained without that theorem. In Gelfand's theory, which deals with commuting observables, one constructs from them a geometric space, on which the wave functions can be defined. The construction of this space requires then some form of Zorn's lemma (about partially ordered sets). But both the Hahn-Banach theorem and Zorn's lemma rest explicitly on the axiom of choice (as a matter of fact, Zorn's lemma is equivalent to it). I am not sure

---

[4] Physicists often think of the commutative algebra as generated by a maximal set of commuting observables (a notion that was introduced by Dirac) and of the geometrical space as the classical configuration space of the underlying physical system: the geometry of the configuration space is therefore derived from the algebraic properties of the observables, and that is really a fascinating result.

whether Gelfand's construction needs a strong or a rather weak form of the axiom of choice, but Von Neumann's work, which is concerned with more general physical systems, definitely requires the strongest form of this axiom.

This conclusion is important for physism. It means that one gains a deeper understanding of the foundations of quantum mechanics when one can rely on the axiom of choice. Said otherwise, this axiom definitely belongs to the language of physical reality, at least as we understand it at present.

# Virtual Processes

**W**hen discussing Feynman histories, we said that they mean that quantum mechanics deals with possibilities, and not with a unique, sharp, reality. We encountered afterward other formulations of the theory, however, and this essential feature was not so obvious with them. The question therefore arises whether possibilities, or virtual events, are intrinsic to the quantum world or whether they are just a feature of a special approach. This question is considered in the present chapter. It deals with the so-called virtual effects, which are encountered in quantum electrodynamics and in the standard model of quarks and leptons. Freeman Dyson (improving on previous work by Julian Schwinger and Shen-Itiro Tomonaga) derived the rules of quantum electrodynamics as we will use it from the Hilbert space formulation of quantum theory, whereas Feynman himself had derived them from Feynman histories. The conclusions one may draw from that important part of physics do not depend therefore on a specific approach and, as might be expected, they confirm the initial statement: at a fundamental level, a grand game of possibilities is intrinsic to quantum reality.

This chapter will not, however, add anything new to the philosophical conclusions on the character of physical laws that will be stated in part 3 and are essential for physism. The present considerations stem therefore only from a desire for dialectical rigor (if such rigor exists) and, since they involve a few technical developments in physics, they can be skipped by readers who feel willing to accept the conclusion.

## Quantum Electrodynamics

Electrodynamics is concerned with the interaction of charges with an electromagnetic field. In quantum electrodynamics, the charges are carried by particles, for instance electrons and protons; rather than the electromagnetic field, one prefers to consider its constitutive particles: photons. Photons travel at the velocity of light, and one must use the theory of special relativity when describing them; for consistency, one must also describe the charged particle motion according to relativity. That cannot be done in a quantum way without considering antiparticles on the same footing as particles and, as we shall see, many electromagnetic processes involve a consideration of positrons (antielectrons).

The number of electromagnetic phenomena is so tremendous that it will be better to concentrate on two paradigms: Compton scattering and electron-proton interaction. A typical experiment in Compton scattering consists first in sending a photon against an electron; the experimental devices are such that one knows the momentum (and therefore the energy[1]) of each particle. One is interested in the process where the photon scatters on the electron, and some detectors then measure the momenta of the outgoing electron and photon. Compton scattering is a paradigm of many processes occurring in ordinary circumstances: for instance, the reflection of light on a mirror or the scattering of the sunlight on the atoms in the atmosphere (which is why the sky is blue).

---

[1] In relativity theory, the energy of a particle is a well-defined function of its momentum and mass.

Electron-proton interactions have two main consequences. An electron can scatter, for instance, on a proton or a
heavier charged particle (that is the origin of resistance in con-
ductors and an important phenomenon in stars). The electron
and proton can also bind together to produce a hydrogen
atom, and the detailed study of the atom is also a topic of
quantum electrodynamics.

## Free Particles

Let us suppose to begin with that the particles (for instance,
electrons) have no charge. They do not interact together nor do
they interact with photons. The electrons and the photons behave as free particles, and we are very near again to our familiar chessboard game. Some refinements must be introduced,
however. Space is, of course, three dimensional; we will not
bother with squares (or rather cubes) paving space and deal directly with amplitudes for going from some space point $x$ at a
time $t$ to another point $x'$ at a time $t'$. It is convenient, in the
framework of relativity, to bring together space and time and
to denote, for instance, the couple $(x, t)$ by a unique symbol $x$
representing a point in space-time.

It is extremely convenient to represent the transition of
a particle from point $x$ to point $x'$ by a straight line joining the
two points (see figure 8.1a). This line has no realistic physical
meaning; it does not represent, for instance, a Feynman history (the line would not be straight and it would not be
unique). It is only a graphical convenience for representing
the quantum amplitude for the particle to go from $x$ to $x'$, and
nothing else. In order to distinguish particles, an electron is
represented by a solid line and a photon by a wiggly line. A
less trivial refinement consists in using the same symbol of a
solid line for an electron and for a positron. They are distinguished by an arrow, pointing either upward (for an electron)
or downward (for a positron). This convention may look

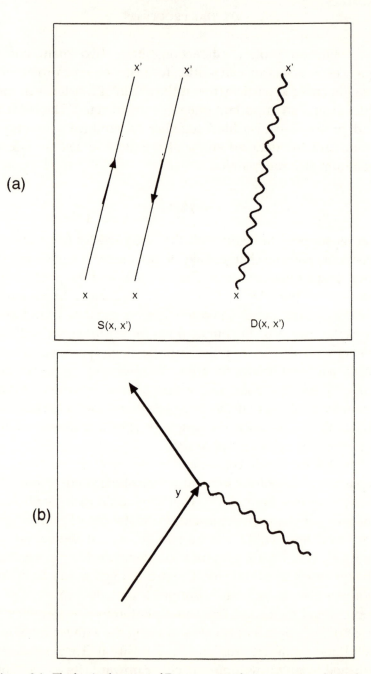

Figure 8.1. The basic elements of Feynman graphs in quantum electrodynamics. (a) Propagation of the space-time point $x$ to the point $x'$ for an electron, a positron, and a photon, respectively. (b) An interaction vertex.

mysterious and its origin lies far down in the nature of quantum relativistic particles, but this book is not a treatise on physics, and we shall simply allow the arrow description as a convenience (an occasional convenience in our case).

The quantum amplitude for the photon going from $x$ to $x'$ is denoted by $D(x, x')$, and by $S(x, x')$ in the case of the electron. The computation of these two functions is just a matter of playing a chessboard game and they are known explicitly (we won't write them, of course). A slight complication is due to the fact that the particles have spin (spin $1/2$ for the electron and 1 for the photon), so that one must use matrices and not ordinary functions; but who really cares, except people who compute?

## THE FEYNMAN RULES

Feynman wrote down the rules for quantum electrodynamics in the spirit of his histories, but Dyson also derived them from the Hilbert space formalism and a previous less transparent construction by Schwinger. An elementary interaction between an electron and a photon at a space-time point $y$ is represented by a vertex as in figure 8.1b. The photon disappears (if it comes from the past, below) or is created (if it goes toward the future, upward). The electron goes on: it cannot disappear, because the electric charge would also disappear and the charge must be conserved.

A quantum process is then represented by a set of graphs describing every possible sequence of free propagations (lines) and interactions (vertices). Figure 8.2a represents, for instance, a graph contributing to the Compton effect. The incoming photon and electron interact at point $y$, where the incoming photon disappears, the electron afterward propagates freely to a point $y'$, where it emits the final photon. This graph is meant only as an illustration of a calculation. Every electron line, i.e., the incoming, the intermediate, and the outgoing ones,

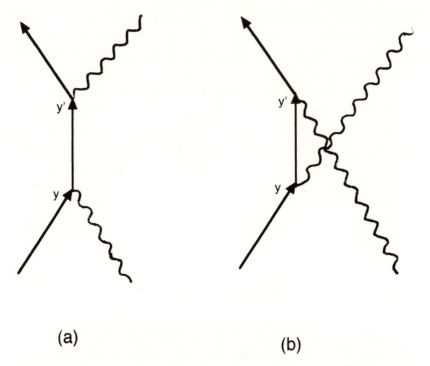

(a)                                    (b)

Figure 8.2. Two graphs contributing to the Compton effect
(electron-photon scattering).

symbolizes a propagation function. The intermediate line, for
instance, is represented by a quantity $S(y, y')$. Every vertex
stands for a factor $e$ (the electron charge) and a matrix accounting for the spins of the interacting particles. One multi-
plies all the quantities that are symbolized by the graph,
propagators $D$ and $S$ for the lines and factors $e$ for the vertices; then one integrates over all space-time for the position
of internal vertices (the points $y$ and $y'$ in the case of the
graph 8.1a). The result is a contribution to the probability
amplitude of the process at hand (Compton scattering in the
present case). The total amplitude is the sum of the contributions of all the possible graphs, of any shape, with any number of vertices.

There are an infinite number of graphs for any process

and the summation of their amplitudes would be impossible if nature had not been kind enough to make the charge of the electron small. Every vertex carries a factor $e$; every couple of vertices therefore carries a factor $e^2$; there are factors involving Planck's constant and the velocity of light in the propagators and they conspire to make $e^2$ occur always in a dimensionless factor $e^2/\hbar c$, which is called the fine structure constant and is equal (approximately) to $1/137$.[2] The nice point is that it is a small number, and more and more complicated graphs with many vertices contribute to the amplitude by higher powers of the fine structure constant, so that the total amplitude appears as a power series in this quantity, and one may retain only a finite number of graphs to compute the probability of the process with a precision good enough for comparing theory and experiment. This technique of summation is called perturbation theory. For instance, there are only two graphs, 8.2a and 8.2b, contributing to Compton scattering at lowest order; in graph 8.2b, the outgoing photon is emitted before the absorption of the incoming photon.

## Virtual Processes

It may be noticed that the final amplitude is a sum of contributions from the various graphs. These contributions are complex numbers, like the amplitude itself, so that they can interfere, constructively, destructively, or something in between. The graphs cannot therefore represent alternative physical processes, but virtual processes acting as so many "possibilities." The question we asked when beginning this chapter therefore receives the expected answer: whatever approach is used for formulating quantum mechanics, it never deals with a unique reality but with interfering virtual "possibilities."

The virtual character of the process in graph 8.2a is

[2] More precisely, $1/137.035999$.

already obvious. It turns out, from the underlying field theory of quantum electrodynamics, that energy-momentum is conserved at a vertex. Suppose then that the two incoming particles have well-defined momenta (and therefore well-defined energy), and the momenta of the outgoing particles are measured. Because of the conservation of energy-momentum at each vertex, the energy and momentum of the intermediate electron can be computed, but it gives a crazy result. From its energy and its momentum, one can compute its mass, but that is not the electron mass! The process does not involve a real, genuine, electron, but a mathematical "thing."

Virtual processes have fascinated physicists and physically minded philosophers. Around an atomic electron, for instance, it looks as if (according to some graphs) there were sometimes pairs of electrons and positrons (with a wrong mass of course). The pairs interact with an electric field as if the vacuum were polarized. No experiment, no logical investigation can give them a reality or a sensible meaning.[3] They give very useful hints, nevertheless, because our imagination works on images, analogies, and it does not require all the features of reality; it is little inspired by a long sequence of algebraic equations, but may guess wonderful possibilities from a mental representation holding something of a familiar reality. The expression "looks as if" rather than "is" would be therefore a safe enough caution for speaking of virtual processes. Unfortunately, it would make sentences too awkward and physicists never do that (except when they write questionable philosophy books such as this one).

We must mention at this point David Bohm's theory. It is a version of quantum mechanics in which the particles are

---

[3] Logical investigation of virtual processes can be carried out by means of Griffiths' histories, which are mentioned in chapter 11. The conclusion is that these processes cannot be described by standard logic: they are nonsense for a logician.

supposed to be "real"; they always have well-defined positions and momenta and they move according to a wave function piloting them. The theory works nicely for nonrelativistic particles, but fifty years of strenuous effort have not yielded a reasonably satisfactory version of it (note that, during that period, the greatest successes of conventional quantum theory were obtained with relativistic systems). Virtual processes would be a headache for people working on that theory: what is real in the positron-electron pairs of vacuum polarization? Surely, they cannot be observed, but the effect itself (i.e., the corresponding graphs) has precise consequences, which have been checked. "It looks as if" a relativistic version of Bohm's theory is impossible, because its existence would conflict with effects negating its basic assumption.

An interesting question is why virtual processes resembling reality occur in calculations. The answer is rather simple: A quantum theory, for instance electrodynamics, rests on definite laws, and these laws are formally simple, even if their consequences can involve a long algebraic elaboration. The quantum field version of electrodynamics, for instance, is given by an equation (extending the "action" we met in chapter 4) no longer than half a line on a page. The Feynman rules come out of it after a typical elaboration of half a book. The consequences of the theory, including the classical ones, are dealt with in series of several books. This means that the basic theory is really a nutshell containing many pumpkins. It is not surprising in these conditions, and even to be expected, that many analogies occur in various places. Vacuum polarization, for instance, evokes an ordinary polarized medium, with positive and negative charges analogous to the quantum positron-electron pairs. The underlying basic theory is simple, unique, and there are consequently strong analogies in the mathematics describing the two situations. The same equations always give the same mathematical consequences, whatever the physical systems one is dealing with, but our minds are swifter with reality than with equations and can jump immediately to their consequences when we are familiar with a realistic example. If not immediately, we can at least guess, get help from images, use imagination in the original sense of the word "image." That's all. Of course, hard labor and verification of the guesses must come afterward, but that is always the fate of intuition in science.

## CHAPTER NINE

# Back to Classical Reality

There is absolutely no doubt that the reality we perceive is classical: unique, continuous, causal, with separate objects well located in space at any time. We must recognize also that the objects we perceive are macroscopic, even when we look at them with a powerful magnifying device. We know that these objects are made of atoms, some very little things that baffle us. Reality, at their level, appears nonunique, it can be discontinuous, it is noncausal, and its constitutive objects, the particles, cannot be distinguished from one another. We rub our eyes and wonder, like Henry the Second about Thomas Beckett: how can a saint come out of a nice drunkard? How can divagating quantum particles turn into the precise motion of a satellite, noncausal laws generate determinism, and discrete bits make up continuity? This is the question to be considered now.

Many fat and thin books have been written on this question (the present author wrote a few middle-sized ones). The titles often mentioned something like "interpretation of quantum mechanics," which was a manner of recognizing that one could not yet write "understanding quantum mechanics." Much has changed in the last few decades under the efforts of many theoreticians and experimentalists, some of the main insights being due, in my opinion, to Murray Gell-Mann, Robert Griffiths, James Hartle, Hans-Dieter Zeh, and

Wojciech Zurek. I will disregard the minor divergences among specialists on the best way of organizing the results consistently and I will consider only their main features.[1]

The present book is in principle devoted to reality and mathematics, not to physics for its own sake. Although most papers on understanding the status of quantum mechanics gave predominance to interpreting, explaining, and now deriving (from first principles[2]) measurement theory, I will not dwell on that subject (except later on for a problem concerning the uniqueness of classical reality). A measurement is a special case of interaction and not the *deus ex machina* it used to be, and its theory is rather technical. I shall try, on the contrary, to stress the weight and meaning of mathematics in the elucidation of the converging quantum and classical realities, with an eye on the later philosophical developments in part 4. This is why the main problems will be approached through Von Neumann's mathematical investigations rather than Bohr's meditations. The great shadow of Hilbert will be there also.

## What Is a Physical Theory?

No aspect of mathematics could leave Hilbert unconcerned and he therefore had definite ideas about theoretical physics, which is so close to mathematics. He was aware of the difficulties that are encountered during a search for the laws of physics and he made no pronouncement on that, but he made clear what should be, in his opinion, the final form of a theory. When he defined the pattern of theoretical physics, his main

---

[1] For more details, one may look at Griffiths (2002) and Omnès (1999b).

[2] These principles were stated briefly in chapter 7, in the section "About the Axiom of Choice."

intent was to disentangle the logical from the empirical questions. He thought, of course, of the axiomatic form of mathematics as the paradigm in matters of consistency. A physical theory has a youth, when it is seeking its way, but when it has reached a mature state, it should rely on a set of axioms: the axioms of the necessary mathematics and some specific axioms stating explicitly the basic concepts of the theory and its fundamental principles. The underlying logic did not reduce exclusively to mathematical logic and it had to be made explicit. When these various assumptions had been stated, the construction of the physical theory was supposed to follow the same course as pure mathematics, proceeding through rigorous proofs to derive the consequences of the principles. An important proposition would become a theorem from the standpoint of mathematics, and a specific physical law in relation to reality.

Some logical questions, already familiar in mathematics, would also occur in theoretical physics. There might be equivalent sets of axioms (think of the equivalence of the formulations of quantum mechanics by Heisenberg, Schrödinger, and Feynman). The question of the independence of axioms would have to be cleared up (for instance, the axioms of measurement theory have been shown recently to be dependent on the basic axioms—or principles—of quantum mechanics: they became theorems). The main characteristic of a mature physical theory would in any case be its purely deductive pattern.

When seen in that light, a physical theory is in a nutshell a branch of mathematics with some specific axioms. It is true, or at least harmonious, if it satisfies two conditions: (i) as a mathematical theory, it is consistent; (ii) it agrees with all relevant experiments, including the new ones it suggests. That is also our definition of a theory in this book (it might have been different in another context, but it fits the present one).

## Von Neumann's approach

Von Neumann had been a student of Hilbert and later his collaborator. He shared Hilbert's views on theoretical physics and the approach of his book, *The Mathematical Foundations of Quantum Mechanics*, makes that clear enough. He introduced the framework of Hilbert spaces, which had a clean mathematical foundation. He stated the superposition principle (which amounts to associating a vector in Hilbert space with the state of a physical system) and introduced observables as self-adjoint operators acting on that space. He performed a great feat of mathematics by developing a rigorous theory of operator spectra. That theory brought him then to the elements of logic adequate for quantum mechanics, and we must say a few words on that.

Let us go back for that purpose to our chessboard game, when we described operators (in chapter 6, the section "The Hilbert Space Formalism"). We dealt then with a complex Euclidean space, which is a simple example of a Hilbert space. The orthogonal projection of a vector $X$ on a normalized vector $Y$ (i.e., a unit vector) is a vector collinear to $Y$ with a measure along $Y$ (algebraic length) equal to the scalar product of $X$ and $Y$:

$$PX = (Y, X)\, Y. \tag{9.1}$$

The orthogonal projection operation has been written here as an operator $P$ acting on the vector $X$ (since it is clearly a linear operation). When the Euclidean space is complex, one can use exactly the same formula.

We saw also that an observable is associated with an operator (or a matrix) $A$, with orthogonal eigenvectors $Y_k$ and corresponding real eigenvalues $a_k$, such that $AY_k = a_k Y_k$. It will be convenient to normalize these eigenvectors. Let us then consider a real interval $\Delta$ containing some eigenvalues of $A$; the corresponding eigenvectors span a subspace $S$ of the

Hilbert space and the projection of an arbitrary vector $X$ on $S$ is given by

$$P(A, \Delta) = \Sigma(Y_k, X)Y_k, \qquad (9.2)$$

the summation being extended to the eigenvectors with eigenvalues in the interval $\Delta$.

What can be said of the projection operator $P(A, \Delta)$? Let us again call it $P$ when there is no risk of confusion. One can easily check that it is linear and self-adjoint. But we know from geometry that the projection on $S$ of the projection of the vector $X$ is again the first projection (since it lies in $S$). That means that $P$ satisfies the algebraic equation $P^2 = P$. Its eigenvalues can therefore be only 0 or 1 (since they satisfy the equation $p_k^2 = p_k$). They are all equal to zero when $\Delta$ contains no eigenvalue and equal to 1 when $\Delta$ is the whole set of real numbers. In the second case, $P$ is equal to the identity operator $I$. One also finds easily that, if $\Delta'$ is the set complementary to $\Delta$, one has

$$Q \equiv P(A, \Delta') = I - P(A, \Delta) \equiv I - P, \qquad (9.3)$$

and

$$Q^2 = Q, QP = PQ = 0. \qquad (9.4)$$

## Propositions

Let us now consider the proposition "the value of the observable $A$ lies in the real set $\Delta$." A very important point is that no reference is made to a measurement; the proposition is just a logical statement. We may associate it with the projection operator $P$ (they are in a one-to-one relation). The logical negation of the proposition amounts to a similar one, namely, "the value of $A$ lies in the set $\Delta'$, complementary to $\Delta$," and it is associated with the projection operator $Q$. The equation $QP = PQ$ implies that the two observables $P$ and $Q$ commute: they are mutually compatible and one can assert them simultaneously. But, then, they exclude each other (since $QP = PQ = 0$). Moreover, one may think of the two numbers 0 and 1 (the eigenvalues of the projection operators) as meaning, respectively, "false" and "true," as in the logic of computers. Lo and behold! Von Neumann had thus discovered a way of describing any physical proposition within the mathematical language of quantum theory and it satisfied the principle of the excluded middle! The road to an axiomatic logic in accordance with quantum reality looked thereby open.

## Doctor Von Neumann's Three Cases

Crash! When he tried to develop this brilliant idea, Von Neumann fell upon three dire problems, which everybody, including himself, for a long time considered insuperable. These problems were as follows.

### Problem 1: Macroscopic Interferences

We mention this problem first because it was certainly the most important one. It appeared only in the last few pages of Von Neumann's book, which were added at the last moment. He there proposed a model of a quantum measurement in which, for instance, an observable $A$ has the two eigenvalues $+1$ and $-1$. The physical system $S$ having that observable $A$ interacts with another system: a pointer sliding along a ruler, its position being denoted by $X$. The pointer was also considered as obeying the rules of quantum mechanics, and Von Neumann devised a clever interaction between the two systems ($S$ and the pointer), which had the following nice properties. The pointer is initially at the position 0 on the ruler, or, more precisely, it has a narrow wave function $\psi(x)$, centered at the value $x = 0$. When the initial state of $S$ is an eigenvector of $A$ with the eigenvalue $+1$ ($-1$), the pointer wave function

110

after the interaction has become $\psi(x - 1)$, ($\psi(x + 1)$), so that it is then centered at the position $x = 1$ ($-1$). This result looked fine at first sight and it seemed to describe correctly a realistic experiment.

But here's the rub. Suppose now that the initial state of $S$ is not an eigenvector of $A$. It is, for instance, the linear combination $|+1\rangle + |-1\rangle$ of the two eigenvectors with respective eigenvalues $+1$ and $-1$. This assumption is perfectly legitimate in view of the superposition principle. When the interaction with the pointer takes place again, the resulting wave function of the pointer turns out to be the sum $\psi(x - 1) + \psi(x + 1)$!

What does this mean? A naive answer would say that the pointer could be close to the position $+1$ or the position $-1$ with equal probabilities. That would look fine. But suppose that, rather than having the pointer slide along the ruler on which the result is read, one proceeds otherwise: a hole is punched in the pointer and the ruler (or a screen) is placed somewhere behind it. What will then be seen when a parallel beam of light falls normally upon the pointer? One would naturally expect that a spot of light will be seen on the ruler, standing either at the position $+1$ or $-1$, with equal probabilities. But no! What quantum theory predicts is that there are conditions under which one can observe interference fringes on the ruler, as if the pointer had been in two positions at the same time.

The result looked crazy of course, but it seemed impossible to deny it, since it rested ultimately on one of the cornerstones of quantum theory: the superposition principle. It was interpreted as meaning an essential difference between the laws of physics at the microscopic and macroscopic levels. Bohr stressed this idea of having two kinds of laws, quantum laws and classical laws, the second ones being necessary for ensuring a meaning to the notion of truth. Heisenberg speculated on a frontier separating the domains of the two types of laws.

The problem had become famous meanwhile, when Schrödinger popularized it with the unwilling contribution of a cat. He considered a system $S$ consisting of a radioactive source, the observable $A$ meaning in that case that there has been a decay, or not, at the time when the measurement is read. The measurement involves a Geiger counter, and we can still think of the pointer on the counter dial, but in place of punching holes, Schrödinger introduced a cat in the measuring device together with a phial of poison that would be broken when the counter registered. He then found that the final state of the cat is a linear superposition of two states, "being alive" and "being dead." This paradox then became known as the "Schrödinger cat problem."

### Problem 2: Classical Statements

When Bohr called attention to the notion of truth in quantum theory, he was raising a question of logic, and we have seen that Von Neumann had found a language for the propositions dealing with a quantum system. Could this language be extended to a classical proposition? A quantum proposition had been associated with a projection operator, but the procedure seemed to fail (at that time it was thought to fail) when one tried to extend it to a classical property. The obstacle was noncommutation. A classical property does not generally involve a unique observable (or several commuting ones), but typically two noncommuting observables. It states, for instance, that a position coordinate lies in a given interval *and* the value of the corresponding momentum simultaneously lies in another interval. Von Neumann could not find a projection operator stating this kind of property and his language appeared therefore strictly limited to the quantum world. It did not look like a universal language.

### Problem 3: A Nonsense

Von Neumann called "elementary predicate" a proposition that can be associated with a projection operator $P$ in Hilbert space. It made sense for any projection, because a projection operator $P$ is an observable and it is also the projection operator corresponding to the proposition "the value of the observable $P$ is equal to +1." Von Neumann thought therefore that any elementary predicate would make sense, but he made two fatal errors. One was to assume that any predicate could be logically linked with any other predicate (as standard propositions can); he thus missed finding Griffiths' histories. His other error was to link the predicates with measurements too narrowly. He could not then avoid the muddle of noncommutation.

He had found a language, but the minimum one should ask of a language is to make sense. If one remembers Hilbert's views, making sense would mean satisfying at least the simplest elements of logic. That condition can be split into the following.

- One should know exactly what is the set of all the propositions. This set is what the logicians call the "field of propositions" or the "universe of discourse" of the theory. We just saw that Von Neumann assumed erroneously that this set included all the possible predicates.
- One must be able to define the negation of a proposition. That condition was satisfied, as we saw.
- One must be able to define the logical connectors "and" and "or" between two propositions; one should also be able to define two logical relations among propositions, namely, the logical equivalence of two propositions $a$ and $b$, as well as the inference relation: $a$ implies $b$. These logical constructions cannot be arbitrary and they must satisfy a definite set of axioms, which is the signature of standard logic.

No matter how he tried, Von Neumann could find neither connectors nor relations satisfying all the axioms of standard logic. He was so baffled by this result that he proposed later with George Birkhoff that the quantum world might perhaps obey a nonstandard form of logic. In a sense, he had given up.

## CHAPTER TEN

# Decoherence

The answers to the first two Von Neumann problems, macroscopic quantum superposition and classical properties, are the topic of this chapter. They are not only interesting for their own sake, i.e., for a better understanding of physics, but they also provide an introduction to a drastic change in the appearance of the physical laws when one goes from the microscopic world to the macroscopic world, from quantum to classical reality. This transition, so radical that it could be called a transmutation, is very important for physism, which could not be a consistent philosophical thesis until that point was cleared up. This consequence, however, will only come out later in this book and we must presently consider the problems only from the standpoint of physics.

## The Von Neumann "Experiment" Again

It came out in problem 1 of the previous chapter that a quantum superposition of two states of a microscopic physical system could be amplified after a measurement into a quantum superposition of two different states of a macroscopic device. Nothing of that sort had been shown experimentally, however, and it would make the world crazy if it happened. Suppose, for instance, that a mother looks at a photograph of her daughter and sees interference fringes in place of the child's nice blue eyes (I take the example of blue eyes because it would be less striking with brown eyes). Or suppose that a mad scientist actually performs Schrödinger's experiment with his neighbor's cat. The animal comes out dead and alive; will a judge condemn the man to a quantum fine, to be paid in quantum money: ten dollars and one thousand, say, though the "and" certainly does not mean \$1010? Many examples, even crazier, could easily be given.

Let us look at the problem from a theoretical standpoint. We certainly will not say that quantum physics plays no role at all in macroscopic objects: it was mentioned, for instance, in chapter 5 (the section "Can One Measure Two Quantities Simultaneously?") that the quantum uncertainty relations are universally valid. They nevertheless have negligible consequences under ordinary conditions, because of the smallness of Planck's constant. We might therefore think of a wave function describing a tennis ball, but that function would imply a very small uncertainty in the position of the center of mass of the ball.[1] Let $\psi_1(x)$ denote that wave function, as it would be, for instance, if the ball had entered a room through one open window, and $\psi_2(x)$ if it entered through another window. We would expect according to quantum mechanics that the ball can enter through both windows simultaneously, so that its wave function would be $\psi_1(x) + \psi_2(x)$. Both components may well be strongly localized, so that it seems that one is dealing with a random process, where the ball could have come through either window with equal probabilities and they are then in two possible distinct positions. But that is not the end of the story. Other measurements (the ball velocity, for instance) can show bizarre

---

[1] As a matter of fact, a macroscopic object is (almost) never "prepared" in a state that would be described by a wave function (one must rather use a so-called "density matrix"). One may still think, however, in terms of wave functions when discussing most questions of principle.

interference effects. The example of a tennis ball is similar to the Von Neumann measurement in the previous chapter; it will be convenient, however, to choose a definite case and we shall consider again the case of a pointer sliding along a ruler, except that both pointer and ruler are now realistic macroscopic objects, made of atoms and not simply mathematical artifacts.

### A Key Mathematical Quantity*

A straightforward mathematical analysis, which we shall not detail, shows that any interference effect depends on the product[2]

$$\psi_1^*(x)\psi_2(x') \tag{10.1}$$

of the two wave functions (actually one of them multiplied by the complex conjugate of the other); the two arguments $x$ and $x'$ are generally different. One can easily compute the interference effects for any observable in terms of that quantity and there would be no quantum interference effect if it vanished; but of course it does not since otherwise one of the two wave functions would vanish.

---

[2] In the case of a two-slit interference device, one recognizes in the quantity (10.1) the product of the wave functions coming from the two slits. One is interested in that case in the intensity of light at a point $x$ and one accordingly takes $x' = x$. If one were observing interferences during the measurement of the ball momentum, one would need to introduce the derivatives of the wave functions (since the momentum observable is given by the differential operator $(\hbar/i)\partial/\partial x$); this derivative itself is obtained by taking the limit of $x'$ tending to $x$. If one were computing interference effects arising from the gravitation potential of the ball (in a state of superposition corresponding to its location in two different places), one would have to introduce a quantity (10.1) involving significantly different values of $x$ and $x'$.

## The Notion of Environment

We mentioned previously that Heisenberg early envisioned that the problem of macroscopic interferences might find an answer if one explicitly considered the atoms in a macroscopic piece of matter. This meant, for instance, that, rather than considering a wave function $\psi_1(x)$ depending on a unique variable $x$ (the pointer position), one would have to consider a much more complicated function $\psi_1(x, y)$ involving again the pointer position $x$ and a huge number of variables, collectively denoted by $y$, denoting the particle positions in the macroscopic object. It should be kept in mind that the number of these variables in a middle-sized object (a mobile phone, for instance) is of the order of a billion of billions of billions, whereas the number of variables $x$ we can actually observe is at most of the order of a few millions or less. These so-called "collective" variables are the ones introduced long ago in classical physics describing the making, shape, mechanism, and motion of the object. This is at least the empirical definition we adopt in a first approach. It happens sometimes that some microscopic variable is actually observed (the point of arrival of a photon on a screen, for instance), in which case it can be included in the $x$'s. The $y$-variables are therefore implicitly defined as the ones we know to exist, but which we do not actually observe or measure.

We do not or we cannot? This is an important question from the standpoint of theory. If we simply did not observe these variables, but we could in principle, we would have to always make them explicit in our mathematical analysis. But if we cannot observe all of them, as a matter of principle, the consequences of the quantum principles must take this fundamental ignorance into account. So, what is the answer? It is: One cannot![3]

---

[3] Rather cogent arguments indicate that, to know the wave function $\psi_1(x, y)$ of an ordinary object as big as a mobile phone, the necessary measuring apparatus would involve tremendously more matter than there is in the observed universe.

The abstract physical system consisting of everything unobserved during an experiment (atoms and electrons inside, molecules in the atmosphere outside, photons in the surrounding light, and so on) is called the "environment" of the "collective" system that is described by the $x$ variables. It can involve an external environment, such as the photons allowing us to see it, or an inside environment (using a somewhat odd association of words), representing the "atomic bowels" of matter.

But then, something has changed in the theory of macroscopic interferences. These interferences are observed on the $x$ collective variables, or other similar variables (momenta, gravitation potentials, and a host of other quantities could be observed in principle); the $y$ variable cannot be observed in any way. This means that in place of the quantity (10.1), the macroscopic interference effects are described by the quantity

$$\int \psi_1^*(x, y) \psi_2(x', y) dy. \qquad (10.2)$$

The integral sign means a summation over the unobservable environment and the situation is obviously changed in a radical way: is it not highly likely that this sum must vanish?

### The Statistical Approach

If one thinks of a pointer actually sliding along a ruler, there must be some friction during the motion when the pointer starts to move. Many atoms are disturbed in the process, many electrons are also perturbed; they are not the same particles in the same place when the pointer moves toward the right or toward the left. The two functions $\psi_1(x, y)$ and $\psi_2(x, y)$ represent therefore two very different quantum states, and the integral (10.2) is expected to vanish. This argument was worked out in mathematical detail during the decade 1950–1960 and it gave the expected result[4]: one does not expect macroscopic interferences.

## DECOHERENCE

The friction approach was not completely satisfactory, however. It could yield an explanation for the disappearance of macroscopic superpositions, but it also left many questions unanswered: the process was too slow to explain every experiment; it did not give a hint why the macroscopic object behaves classically; it was essentially an explanation by randomness, which did not account for some of the deepest qualitative differences between quantum and classical reality.[5]

Rather than considering the displacement of atoms under friction, one must look at the complete wave function $\psi(x, y)$ (equal in our example to $\psi_1 + \psi_2$) and consider its phase. This phase depends on $x$ and $y$ and it varies rapidly. This variability has several causes: the scattering of atoms and/or electrons introducing scattering phase shifts, a marked instability of the environment under its perturbation by a slight collective change in position or velocity,[6] chemical or electromagnetic

---

[4] N. Van Kampen (1954), relying on an ergodic quantum theory by L. Van Hove; A. Daneri, A. Loinger, and G. M. Prosperi (1962).

[5] I did not sufficiently appreciate the limitations of the ergodic quantum theory and its application to macroscopic quantum superposition when I sketched the history of decoherence in *Understanding Quantum Mechanics*. The importance of Hans-Dieter Zeh's insights was not then stressed well enough and I am glad to seize the present opportunity for doing it and correcting a misleading judgment.

[6] The energy eigenvalues of the environment Hamiltonian are extremely close together, and perturbation theory indicates therefore a high phase sensitivity (the famous "small denominators" instability).

changes of state of a particle, photon escapes, and so on. All these mechanisms are much more rapid than the displacement of atoms associated with friction, although they have the same formal origin in the interaction of the collective subsystem with the environment. A loss of phase correlation is much quicker than an energy loss through dissipation, and this *decoherence* effect is therefore a much more efficient mechanism for the destruction of quantum phase coherence (Zeh 1970).

### *The Theory of Decoherence*\*

The phase effect we described is certainly the fundamental explanation of the decoherence effect destroying macroscopic interferences. Its origin is rather intuitive. One also refers sometime to more abstract arguments, which may look intuitive to a mathematician: when, for instance, one takes two different vectors in a Hilbert space huge enough to represent billions of billions of particles, the chance that they are practically orthogonal is very high. There is, however, nothing more evasive to calculation and challenging for theorists than the phase of a wave function involving many particles. People accordingly used various models, more or less representative of known situations. A useful theory was also obtained by adapting the quantum theory of irreversible processes: decoherence is undoubtedly an irreversible process, although its relation with entropy increase is subtle.[7]

Many different cases can be envisioned and we had better concentrate on the most frequent one in ordinary conditions. One considers the "reduced matrix density," which is defined by the quantity

$$\rho(x, x'; t) \int \psi^*(x, y; t)\psi(x', y; t)dy. \tag{10.3}$$

---

[7] This approach, which was developed by Omnès (2002), was suggested by Roger Balian.

It involves the total wave function and not the two components of a superposition as in the quantities (10.1) and (10.2), because decoherence is a universal process. It is not restricted to the Von Neumann–Schrödinger superposed wave functions in a quantum measurement. We have also introduced time explicitly, because decoherence is a dynamical process and one is very interested in its actual rate.

In many cases, the "matrix" (10.3) tends to become approximately diagonal.[8] Except very near the diagonal (corresponding to $x = x'$) the "matrix element" $\rho(x, x')$ decreases exponentially with time, typically as $\exp[-\mu(x - x')^2 t]$. The "decoherence coefficient" in the exponent is very large. To give an idea, let us consider the academic case of a Von Neumann pointer with mass $M$ at a rather high temperature,[9] which is denoted by $T$. Let $\tau$ denote the dissipation time, i.e., grossly the time at which the pointer has lost one-half of its kinetic energy under the effect of friction, when moving freely on the ruler. One has then the relation

$$\mu \approx \frac{MkT}{\hbar^2} \tau^{-1} \tag{10.4}$$

($k$ is as usual the Boltzmann constant).

This is certainly not the right place for discussing the details of decoherence, but one may notice a few significant features in these rough formulas: The square of the very small Planck constant in the denominator indicates that decoherence is an extremely efficient and rapid effect (as a matter of fact, it is by far the most efficient effect at a macroscopic scale). The factor $\tau^{-1}$ is governed by dissipation; it appears because both decoherence and dissipation arise from the same collective-environment

---

[8] I call this quantity a matrix as many physicists do; a mathematician would rather consider it as a linear functional over the set of observables.

[9] Without entering into technicalities, one can say that the temperature is high enough for the quantum effects to be negligible in thermodynamics.

interaction (which is responsible for the energy transfer from the collective motion to the thermal motion of atoms). Putting numbers in the formulas, one finds, however, that decoherence is tremendously more rapid than dissipation.[10]

## *Observing Decoherence*

For a rather long time, several models yielded trustworthy predictions on the existence, the main characteristics, and the rate of decoherence. But the only sanction a physicist respects, namely, the result of a clean experiment, was still lacking. The reason for this strange situation contained something of a paradox: the effect could not be observed because it was too efficient! One could not catch it when it was acting. Before any action or any measurement had been made, decoherence had already done its day's work and left no trace of quantum superposition.

The effect was finally observed a few years ago (Brune et al. 1996). The system undergoing decoherence was a very large excited rubidium atom (in a so-called Rydberg state with a very large radius) and the environment consisted of

---

[10] One may notice that decoherence does not act when $\tau^{-1} = 0$, i.e., when there is no dissipation. In the case of light interferences, where there is obviously no decoherence, photon-photon interactions are extremely weak, dissipation is completely negligible, and $\tau^{-1} = 0$. Some clever superconductive systems also show very little decoherence and accordingly exhibit a quantum behavior although they are macroscopic. One should not therefore consider that "macroscopic" is universally equivalent to "classical."

Although frequent and even typical, the tendency of $\rho(x, x')$ to become almost diagonal in a definite "coordinate basis" (also called a "pointer basis") remains a matter of controversy. Zurek (2003) proposed a conjecture, based on a variety of models, implying always this kind of diagonalization, although sometimes on a nontrivial basis. Omnès (2002) used a rather general theory and found mathematical cases where diagonalization does not hold. The practical consequences of decoherence remained valid, however.

a resonant electromagnetic field in a cavity. Increasing the number $n$ of photons in the cavity was tantamount to increasing the number of degrees of freedom in the environment. One could then see some well-understood quantum interference patterns disappear gradually when $n$ increased from 0 to 10, confirming the existence of the effect. Since, moreover, the theory could be trusted in that case, it could be compared with the experimental data and the agreement was excellent. No doubt therefore: the decoherence effect exists!

## CLASSICAL BEHAVIOR*

Decoherence suppresses quantum interferences at a macroscopic level. Furthermore, once decoherence has taken place, the system behaves classically in most cases. Moreover, the existence of decoherence has been confirmed recently by experiment. Such are, in a nutshell, the main conclusions of this chapter. The comments I shall now add may answer some further questions of knowledgeable readers, but they are not essential and may be skipped without inconvenience.

Decoherence has much in common with a case of destructive interference: many inaccessible phase differences in the wave function of the environment conspire to erase the quantum phase correlations at a macroscopic level. One can compare it with the case, already considered in chapter 4 (the section "Wave Functions"), when an electromagnetic wave enters into a transparent medium; all the atoms emit small wavelets, which interfere destructively almost everywhere, except along a unique direction where they build up constructively a refracted wave moving with a velocity smaller than in vacuum. This is the best analogy between a familiar physical phenomenon and decoherence, but it cannot show the most important characteristic of the effect, which is the suppression of superposition effects. Optics remains linear; it preserves the possibility of interferences inside a transparent medium as well as in vacuum. Decoherence breaks linearity, and that is certainly its most important consequence.

It does not mean of course that linearity as a fundamental property of quantum mechanics (through the superposition principle) is perturbed and would suffer correction by new effects that were neglected. Nothing was neglected and the quantum principles remain the impervious and universal form of the laws of physics. The difference is elsewhere, in the other basic principle of quantum physics: noncommutation. In optics, as always in classical physics, physical quantities are numbers and linearity means an addition property of classical electromagnetic fields: numbers. The quantum superposition principle holds for wave functions, whereas noncommutation implies that the statistical data for an observable are quadratic in the wave function. It should not therefore be surprising that the actual disappearance of superposition goes together with a return to commutation. In other words, a macroscopic system behaves generally in a classical way after undergoing decoherence.

No hint of a more precise proof will be given here however, because it would rely in one way or another on a beautiful mathematical theory embedding elementary quantum mechanics and classical physics into a common framework. This theory, which is known under the names of microlocal analysis or pseudo differential calculus, was developed mainly around the 1970s and it is highly nontrivial. It had two precursors in physics: the transition from a density matrix to a classical probability distribution in position and momentum by Eugene Wigner in 1932 and the related transition from quantum observables to classical dynamical variables (which are functions of $x$ and $p$) by Hermann Weyl in 1950. They both turned out to play a role in the "correspondence" between quantum and classical physics: Wigner's distribution as the simplest link relating decoherence with a classical behavior and Weyl's calculus to show how noncommutation disappears after decoherence.

We mentioned in the previous chapter that one of the problems impeding Von Neumann was the apparent lack of a projection operator expressing a classical property. This "problem 2" was given an answer by the founders of microlocal analysis, and particularly Lars Hörmander, around 1970. We shall not develop it, except for mentioning that it refers to a domain of values for position and momentum far exceeding the constraints of the uncertainty relations; in other words, the motion of an electron in a television tube, for instance, is well defined classically as long as no detail $\Delta x$ and $\Delta p$ of position and momentum is such that the product $\Delta x \cdot \Delta p$ becomes comparable to $h$.

When one considers accordingly that the fundamental laws are quantum laws and the classical ones are their macroscopic consequence, a classical property can still belong to the language of quantum mechanics (which relies on projection operators); such a property can remain true as long as decoherence disentangled it previously from alternative properties. Or, said otherwise, decoherence is essential for understanding why quantum reality can show sharpness, in the sense of distinctive events with no interference of any kind; the contact with quantum foundations—including the language of projections—nevertheless remains valid. Said again in another way, Von Neumann's language was indeed universal and it can express everything occurring in physics, but its extension to classical reality rests on decoherence.

## CHAPTER ELEVEN

# Did You Say "Paradox"?

**I** think I can say safely that nobody understands quantum mechanics."[1] Feynman wrote this famous sentence forty years ago and nobody contradicted him. If nothing had changed in the meantime, there would be no point in wondering now about the nature of mathematics (which apparently nobody understands) and trying to relate it to physical reality. Has there been such a change? Opinions are still conflicting but I will try to give some arguments in favor of this point of view in the present chapter.

## A Paradigm

According also to Feynman, the paradigm of every paradox in quantum mechanics is the two-slit interference experiment (which we met in chapter 4, the section "Feynman Histories").[2] To show how truly absurd this experiment is, suppose a law-abiding citizen complains of a mishap he suffered to a policeman and the dialogue goes as follows: The thing entered through the windows and hit my photograph hanging on the

[1] Feynman 1965, chap. 6, 129.
[2] Feynman, Leighton, and Sands 1965, vol. 3, chap. 1.

opposite wall, right on the nose!—What thing?—Let's say some sort of wave, like a sound wave.—But how could a wave hit your picture on the nose?—Let's say it hit like a bullet and made a hole in the wall.—So, it was a bullet or something like that. O.K., if you know through which window it entered, we might find who fired it.—Well, I'm a physicist and an expert on waves. I told you already that the thing entered through the windows: there are two of them and I mean both windows, at the same time. Otherwise, the thing would never have got my photograph, which does not hang in an exposed position.—You may be an expert, perhaps, but you're crazy, for sure!

Perhaps we might try speaking another language, a foreign language in which the story would not look so absurd. The most convenient one is undoubtedly Von Neumann's language where a projection operator stands for a sentence. There is then no problem in expressing the fact that the thing—a particle in that case—hit the wall at some point. Drawing a net of squares on the wall as in our chessboard game, every statement such as "the thing hits the square number $n$ on the wall" can be expressed by a projection operator $P_n$. This operator can also specify at what time $t$ the thing hit the wall.[3] When saying that "the thing entered through both windows," we may think of a thin region of space $R$ consisting of two parts, each one bounded by the frame of one window, and write down the projection operator $P_R$ expressing that the position $X$ of the thing is in $R$ at some time $t'$.

We may then introduce the initial wave function $\psi$, expressing how the thing, particle or wave, was prepared. It seems reasonable to consider the quantity

$$|P_n(t)P_R(t')\psi|^2 \tag{11.1}$$

---

[3] In Heisenberg's formulation of quantum dynamics, the observables evolve in time and the projection operator is considered as an observable.

as giving the probability of the "history" according to which "the position of the thing was in region $R$ inside the window frames at time $t'$ and hit the square number $n$ on the wall at time $t$." More generally, a sequence of quantum events, occurring at successive times and expressed by projection operators, will be called a history, and some sound axiomatic arguments show that the quantity (11.1) is the only sensible candidate for its probability; the arguments do not ensure, however, that this quantity satisfies all the axioms of probability theory.[4]

Have we gained something? Not much apparently, except that the notion of quantum events does not appear to be restricted to the special occasion of a measurement, as it used to be in Bohr's interpretation. The real gain appears when we react like the policeman and try checking whether the thing hitting a square on a wall can pass through one, only one, of the two open windows. For that purpose we split the region $R$ into two regions $R_1$ and $R_2$ each covering one window; we introduce the corresponding projection operators $P_1$ and $P_2$ (noticing that $P_R = P_1 + P_2$) and two sets of histories, stating, respectively, that "the position of the thing was in region $R_1$ ($R_2$) across the first (second) window at time $t'$ and it hit the square number $n$ on the wall at time $t$." The corresponding probabilities are defined as the quantity (11.1), replacing $P_R$, respectively, by $P_1$ and $P_2$.

Thus we find the origin of the paradox of the two windows. A straightforward computation shows that

$$| P_n(t)P_R(t')\psi |^2 \neq | P_n(t)P_1(t')\psi |^2 + | P_n(t)P_2(t')\psi |^2, \tag{11.2}$$

which means that the probabilities are not additive: the probability of crossing both windows is not the sum of the probabilities of crossing each one of them! As a matter of fact, we were not dealing with sensible probabilities!

---

[4] These "Griffiths histories" should not be confused with Feynman histories.

A mathematician would say that there is nothing surprising in that result: the amplitudes are additive (since $P_R = P_1 + P_2$), but their squares are not, at least generally. Griffiths' idea of consistent histories was to investigate the significant cases when these two conflicting addition properties turn out to be consistent.

## CONSISTENT HISTORIES

The previous construction provides examples of the "histories" that were introduced by Robert Griffiths in 1984 to describe a series of quantum events. Every event is associated with a projection operator (specifying the time when it occurred); a history amplitude is obtained by letting these operators act on the initial wave function in their order of occurrence; the probability of the history is defined as the square of this amplitude. Since quantum events are random, a unique history does not describe the various events that can happen in a given situation or during a specific experiment, and one must introduce alternative histories (stating, for instance, that the thing had hit a square $n'$ rather than the square $n$). One thus obtains a family of histories, but their probabilities are not generally additive.

Griffiths defined "consistent histories" as a family of alternative histories having additive probabilities. They are then genuine probabilities, as a mathematician would have them, satisfying all the basic axioms of probability calculus; they are no longer toy quantities that a careless physicist chose to call "probabilities" for unsubstantiated reasons. Griffiths wrote down the equations (involving the projection operators and the initial wave function) ensuring the required property of addition and he called them "consistency conditions."

Many cases have been worked out, showing the great value of the idea as a paradox-killer. Considering again the example of the two-slit experiment, the paradox amounts to a comparison of two different families of histories:

- Family I contains a history stating that the particle position was inside the region of the two slits at some time $t$, then the wave function was a sum of two waves originating from the two slits at a time $t'$ (the corresponding projection operator consists in projecting on the Hilbert space vector corresponding to this wave function); finally, the particle hit the screen in some region $n$ at time $t''$. All the possible values of $n$ are considered, of course, and the family is completed by alternative histories; the simplest way to complete it consists in considering the particles that could not cross the slits and the wave functions orthogonal to the correct one. When the times $t, t'$, and $t''$ are well chosen (corresponding to a quantum unitary evolution between times $t$ and $t'$ and then between $t'$ and $t''$), the consistency conditions are satisfied and the family of histories makes sense.
- Family II is identical with family I, except that it considers as alternative possibilities that the particle crosses either one slit or the other at time $t$, and the wave function is centered at one slit or the other at time $t'$. The consistency conditions cannot then be satisfied and these histories cannot make sense.

The question of what it means to "make sense" or not (or the meaning of "consistency," when speaking of consistent histories) was made clearer by the present author: It turns out that the consistency conditions, which are necessary for the existence of genuine probabilities, imply also the validity of standard logic in the framework of a consistent family. One can define the negation of a statement (consisting of predicates belonging to a subset of histories), one can also make explicit the logical connections "and" and "or" among such propositions, and finally one can give unambiguous rules for the two relations "$a$ is logically equivalent to $b$" and "$a$ implies $b$." The basic axioms of standard logic were automatically satisfied

130

and the result therefore provided an answer to Von Neumann's problem 3.

What was most impressive when consistent histories were proposed was that they said nothing of measurements; the screen on which the particle arrived was not supposed, for instance, to allow detection. The propositions describing the quantum events were found to make sense, or not, in a nontrivial way on the sole basis of logic. Furthermore, it was not an exotic logic, especially cooked up for quantum purposes, but good old standard, classical, and Aristotelian logic.

## THE MULTIPLICITY OF QUANTUM FRAMEWORKS AND THE UNIQUENESS OF CLASSICAL REALITY

One may reasonably consider that one understands quantum mechanics when one has at one's disposal a reliable logical method, which avoids paradoxes and gets rid easily of subtle pitfalls (like tempting us to believe that a particle cannot cross two slits together). One must be aware, however, of the fact that there often exist different families of histories that can account for the same experiment and are mutually incompatible. It happens, for instance, when one tries to imagine what is the position, or what is the momentum, of some particle at a given time. It may happen that these considerations lead to consistent histories in both cases, though the two of them are, of course, mutually exclusive. Griffiths proposed the name of "quantum frameworks" to denote a definite consistent family of histories, within which standard logic applies rigorously. The reason for their multiplicity is always the same, namely, that one cannot beat noncommutation. As a matter of fact, the existence of incompatible quantum frameworks is only another—more precise—version of Bohr's principle of complementarity.

Some people are reluctant to accept this multiplicity of logically consistent descriptions of the quantum world. They supposed presumably that a clarification of quantum logic would bring us back to some simple form of realism, or they had no occasion to ponder the existence of different fields of propositions—also called "universes of discourse"—in mathematical logic (the universe of discourse of topology, for instance, is completely foreign to the universe of discourse of algebra). This is anyway a very old question, which goes back to 1928, when Bohr introduced the idea of complementarity, and Bohr's well-known method for getting rid of it remains perfectly valid.

In Bohr's approach, the method for avoiding the problems from complementarity consisted in considering only classical macroscopic events when describing an experiment or some natural event. That simple proposal, which amounts to going back to classical reality, remains perfectly possible, even when one wants to rely explicitly on the quantum principles. There are two kinds of classical events worth mentioning in a description: ordinary classical ones, and the results of a quantum measurement. We already said in the previous chapter (Section 4) that one can describe ordinary classical events in a quantum language by using Hörmander's projection operators. This is valid also for the result of a measurement, or any other classical amplification of a quantum event. There are two essential points in this reduction to actual observation. (i) The resulting description is unique, with no remains of the ambiguities resulting from complementarity. One might then say that the uniqueness of classical reality is recovered when one sticks to macroscopic events. (ii) The corresponding family of—classical—histories is consistent. It should be stressed that the effect of decoherence during a quantum measurement, together with the ensuing classical behavior of the measuring devices, is essential when consistency is established (through a check of the consistency conditions). Decoherence appears once again in this case as the midwife of classicality.

## Hunting Paradoxes

We have mentioned several times that paradoxes haunted quantum mechanics before the discovery of decoherence and consistent histories. Some of them arose from the assumption of wave function collapse after a measurement. Frequent paradoxes came from smuggling some rules of classical reasoning into the quantum domain where they do not belong; they gave rise to logical inconsistencies. Many others resulted from mixing the results of two incompatible quantum frameworks. Some paradoxes were raised for very serious reasons, when the consistency of quantum physics was still in doubt; the Einstein-Podolsky-Rosen example is the main one of that kind. Some people also enjoyed proposing startling new ones as a kind of sport, not very different from crossword puzzles (since they also relied on the ambiguities of language).

But there is no real paradox in quantum mechanics. Griffiths (2002) has devoted a significant part of a recent superb book (*Consistent Quantum Physics*) to a thorough discussion of the most famous ones which were supposed to occur at one time or another in quantum physics. I will give only a short list of them and refer to Griffiths' book for their elucidation; as a matter of fact, I believe that only a few historians of physics will remember them within a few decades, just as the Richard and Burali-Forti paradoxes in the theory of sets have now become the private property of the historians of mathematics.

Here is the list. There are counterfactual paradoxes arising from the question "What would have happened if some state of affairs had been different from what it actually was?" (or, as one says in circles where counterfactuals are not highly considered: would a kangaroo topple if it had no tail?). The main point, when dispelling such a paradox, consists in showing that it involves a combination of several incompatible quantum frames and therefore violates logic. There are also

133

delayed-choice experiments, which cease to be problematic as soon as the result of a measurement is expressed by a projection operator and not considered as the effect of a wave function collapse.[5] There are the so-called interaction-free measurements, such as the Elitzur-Vaidman (1993) bomb-testing paradox, which I will not describe. They may look surprising as long as one does not realize that a lack of direct interaction between two objects does not preclude a logical link between their properties, going back to a common preparation. There is an enormous literature concerning the Einstein-Podolsky-Rosen paradox, which belongs to this category and into which I cannot enter[6]; its main outcome is to have yielded a proof of the objective existence of entangled states, although it boils down simply to the existence of wave functions for several particles that are not products of one-particle wave functions. Aharonov and Vaidman's three-box paradox is a case of mixing incompatible quantum frameworks, like the subtle Hardy paradox, an amusing surprise in a domain where one had otherwise become surfeited (Aharonov and Vaidman 1991; Hardy 1992; see also Greenberger et al. 1990).

[5] In a delayed-choice experiment, one decides at the last moment, when a photon has already entered an interferometer for instance, whether or not one will activate a detector located in one of the arms of the apparatus. This situation was discussed by Wheeler (1978, 1983). An experiment showing that the result is the same as if the detector had been permanently activated—or not—was first made by Alley et al. (1983).

[6] The most important references are Einstein et al. (1935), Bohm (1951), Bell (1964), and Aspect et al. (1981).

# PART THREE

## THE CHARACTER OF PHYSICAL LAWS

We are now ready to enter into the main topic of this book: a proposal concerning the nature of mathematics, according to which the substance of mathematics belongs to the laws of nature. We call it physism because, among all the forms of science, physics stands as the one in which the laws suggest this simple idea most directly. This name should not be misunderstood, of course, as a reduction of mathematics to the scientific discipline of physics, and the Greek word *phusis*, meaning "Nature," should be heard in it. The proposal itself is not obvious although it can be defined in a few words, and it has practically disappeared from the philosophy of mathematics for the last century or so. We saw why already: because of the historical discordance between the two great crises of mathematics and physics, and also a long suspicion concerning the interpretation of quanta. If there is something right in the idea and if this book contains a spark of truth, it will mean that this time is over and one can harvest many insights that appeared during the last century in our knowledge of the laws.

The study of these insights will proceed in two steps, corresponding to parts 3 and 4 of this book. The present one will try to cast a "philosophical" look on the laws of nature, as we know them presently, while shedding some light on their close relation with mathematics. The fourth part will take the reverse approach and show the meaning one can give to the words "substance" and "belonging" when saying that the substance of mathematics belongs to the laws of nature.

But what does one mean, when speaking of the laws of nature? And what is nature (or "reality" as it is now often called)? These are tricky questions, and I believe an important key to the problem we consider is to bypass at least the second one. Everybody is more or less confused about the meaning, height, and depth of reality and we had better recognize the existence of this confusion. We may even be happy with it, because it means that something essential is still unknown and remains open to research and thinking. The question we want to investigate has little to do, fortunately, with the ontological aspects of reality. It does not depend on any form of realism assuming that reality can be known as a matter of principle (what principle?) and asking to what extent it shares the evidence of classical reality. Our question is restricted to the empirical laws governing reality and their relation to mathematics, so that perhaps (and I would even say "certainly") we know enough of the laws and of mathematics to get that point elucidated.

The first step will thus consist in looking at the meaning of the laws. But though the laws themselves are known to a certain degree and have been confirmed by experiments, there is no universal agreement about their meaning, which is a matter of philosophy. My personal experience in this domain has been affected by the rather recent progress in understanding the relation of quantum reality to classical reality, particularly through decoherence. One might discuss whether or not this is a distorted perspective, but I think it is the little grain of salt, the last insight after many others, clearing the way to physism. It is essential anyway for the picture of the laws that will be given here.

This picture is twofold and will be described in two successive chapters. The first one deals with the character of the fundamental laws, namely, the quantum laws governing particles and the laws of space-time. The second chapter will be concerned with the classical laws acting in macroscopic matter. This splitting emphasizes the very different characters of the laws on the two levels, a separation that is due to decoherence.[1]

---

[1] General relativity is often assumed to be a classical version of some wider quantum theory. What this theory should be, however, whether string theory, noncommutative geometry, or something yet unknown, nobody can say with certainty, and the role of

The characters of the laws would become utterly confusing if one ignored this essential difference, and no philosophical implication could be drawn from them.

---

decoherence is somewhat unclear in that case. This is why I preferred to include the characteristics resulting from general relativity in the present chapter.

## CHAPTER TWELVE

# The Character of Fundamental Laws

**W**hen Poincaré reflected on the relations between mathematics and physical reality, he was already aware of some mysteries lurking behind the unassuming evidence of that reality. One hundred years later, we have reached a point where nothing looks obvious in it any longer; "realism" has become an ontological doctrine and a subject of controversy; nobody in a philosophy department confuses "real" with "obvious," in spite of etymology, and Bohr emphasized that one must still learn as if they were new the meaning and the extent of the word "reality." If we intend to go back to the traditional relation between reality and mathematics, we must avoid the philosophical morass surrounding the various meanings, strata and degrees of reality.[1] As a matter of fact, we understand much better the laws of nature than the thing (res) itself, so that a cautious consideration of reality through the character of the laws will be the approach we follow.

*The Character of Physical Laws* was the title of a great book by Feynman, but it can also be understood as a program of philosophical research, and Feynman's book did not put an end to it. This is the sense in which I borrow this title anyway. We noticed previously a deep difference in the expression or

---

[1] These difficult and subtle questions have been carefully analyzed in several books by Bernard d'Espagnat (1995, 2002).

the appearance of the laws at a fundamental quantum level and at a macroscopic one, on the two sides, so to say, of the decoherence edge. This is why this chapter will consider only the fundamental laws, while the next one will take care of their junction with classical reality, beyond decoherence. I suppose it goes without saying that my intent is not to give a thorough discussion of the deep problem of the character of the laws—too seldom evoked, by the way—but only to stress a few points of importance for the philosophy of mathematics.

## The Laws Exist

Something governs the natural phenomena and one usually thinks of that as the "laws," although the underlying idea has been questioned from time to time.[2] The notion of law, which was borrowed from politics and theology, agrees nevertheless with the manner in which most results of research are stated, elaborated, and checked, and there is no doubt, for instance, that the organic rules of quantum mechanics, which governed the growth of physics during the twentieth century and remained in the meantime essentially unchanged, fully deserve the name of laws.

The existence of fundamental and universal laws ruling nature was clearly stated for the first time by Aristotle in his *Lessons on Physics*. One often reproaches him for his prime assumption concerning motion ("there is no motion without a mover"), or one makes fun of some of his explanations (stones fall for the love of earth and flames rise because of a desire for the sky), but the fact that a human being envisioned universal laws when he did not know any opens a fascinating vista on the human mind and its intuition. Anyway, Aristotle's considerations were clearly a matter of ontology and we cannot follow him on that ground.

[2] See, for instance, Kuhn (1962) and Van Fraassen (1990).

Ontology would rather be an impediment in the present case, because asserting the existence of laws as something unassailable is somewhat contrary to the spirit of science. We only know that some laws agree with all the facts we are aware of up to now, with the precision at our disposal and within the domain we can reach. On the other hand, this declaration is too pessimistic if it does not mention the huge number of confirming facts, their very high precision, and the immense domain of modern science.

The history of science provides the strongest incentive for believing in the existence of universal laws. Some of its main steps coincided, moreover, with a stage of unification between previously different fields of knowledge, and they strongly emphasized the unity of the laws. They were, for instance:

- The unification of terrestrial and celestial dynamics with Newton
- The same for terrestrial and stellar physical processes with Kirchhoff
- Electricity and magnetism joining into electromagnetism in the first half of the nineteenth century
- Optics becoming a branch of electrodynamics, after the synthesis of Maxwell's laws
- The principle of inertia embedded into the geometry of space-time, according to General relativity
- Atomic physics and chemistry united through quantum mechanics
- The quantum laws encompassing two lower levels in the structure of matter: nuclei and the quarks inside hadrons
- Electrodynamics and weak interactions united into the common dynamical framework of electroweak interactions
- Electroweak interactions and strong interactions sharing the common mathematical framework of gauge theory

One could add many items to this impressive list, and particularly the fact that no new physical or chemical laws have been found necessary for understanding the mechanisms of biology or explaining the content, the extent, and the history of the universe. Teachers do not usually emphasize the unity of the laws, however, because it would lead them to start from the most abstract foundations before strolling toward more familiar surroundings, which is contrary to pedagogy. Although this unity looks deep and obvious, some philosophers have questioned it,[3] and there are also discussions on the best way to express it, under the headings of reductionism, emergence, or complexity theory (Cornwell 1995), but that is not essential.

## Against Pure Empiricism

I believe the modern attitude about the existence of laws is too practical and it makes us almost blind to the marvel of their presence. This shortsightedness has a long tradition in philosophy, from Hume's matter-of-fact attitude toward the regularities of nature to Sartre's insistence on absurdity. People who never themselves entered into the subtleties of a theory or the intricacies of an experiment are tempted to believe that scientific knowledge amounts to no more than straightforward empiricism. That was, after all, Hume's definition of the laws of nature when he made them summaries of a collection of facts; Paul Valéry did not look farther when he described science as the "collection of never-failing recipes." This narrow and too widely shared view of science hampers every possibility of understanding the close relation between experimental empiricism and mathematical theory, and therefore of appreciating the link between mathematics and reality.

Let me illustrate this by a tale, which will hopefully be clearer than a philosophical discourse. Once upon a time, a powerful race existed far away in this galaxy. Its people had a wondrous memory and they could live millions of our

---

[3] See particularly Cartwright (1983).

years. Their species had lasted more than one billion years but, except for that, it was a very ordinary science-fiction population.

Their high capacity for memory had two remarkable consequences. These people were keen observers and they never ceased accumulating new data under a multitude of circumstances. They had, however, no inclination of any sort toward abstraction, since abstracting is essentially forgetting supposedly inessential features of an object, whereas these people could not forget anything. They never invented mathematics therefore, except for the basics of arithmetic allowing them to classify or compare data and also to interpolate between them. Their ability at making long elementary computations (once more a gift of memory) and their knowledge of a vast number of facts could be used for making predictions, including very precise ones. They had thus discovered (or should we say invented?) many a clever appliance through trial and error. They had the most powerful microscopes of every possible type and they traveled everywhere in the galaxy. They had visited the neighborhood of black holes and seen pulsars accelerating particles at incredibly high energies. Verily, they knew every fact we know and much more than that, although their knowledge remained a pure collection of facts. They were perfect empiricists.

The aim of this little tale is only to make clear that a science without the idea of laws is conceivable, and it could be a highly evolved science. The inhabitants of this remote place would have been astonished, however, if they had heard of David Hume saying that the laws of science consist of a summary of facts. They would have considered the statement as highly metaphysical, and their amazement would have been due not to our use of experimental methods, but to our interest in summaries and the possibility of turning them into laws. Our laboratory devices would have looked trivial to them, but the idea of a Pythagorean mathematician who could *prove* that the diagonal of a square is an irrational number would have seemed to them sheer nonsense, or perhaps the deed of a god.[4]

One of the greatest mysteries of science is the moment of its conception, when adequate concepts are conceived as the best expression of phenomena, and a law is conceived as the right organization between these concepts. It is the moment when the fall of an apple shows acceleration as the effect of a force through a newly conceived calculus. It is Maxwell's slow groping through intricate models of moving charges to a few pure equations, Heisenberg's incredible noncommuting numbers, and many other more recent examples. It is the creativity of the human mind, the same when Feynman sees the cooperation of histories and when Riemann projects geometry toward new horizons.

## The Laws Are Consistent

Karl Popper has recounted the strong impression he felt when learning of Eddington's measurement of the curvature of light rays in the vicinity of the sun. The prediction that light rays could be curved was one of the most striking consequences of general relativity, and its experimental test was a question of life or death for the theory. Popper's reflections on the event led him, as is well known, to the notion of "falsification" as the central test of truth in science. There is something dramatic in the idea of falsification, particularly impressive in French where the two words meaning "false" and "scythe" are written in exactly the same way (*faux*). As a matter of fact, Popper was not much known for a long time among scientists and, when some great test of a theory was communicated during some international meeting, there was little sense of drama

---

[4] This apologue occurred in a novel (*L'espion d'Ici*) by the present author (2000).

among the audience, no feeling that the theory had been salvaged from a great danger, but rather the comfortable fulfillment of an expectation. Many scientists had already scrutinized the theory under test and it was widely expected to be right. This feeling of expectancy is certainly more representative of the reality of research than the famous scythe of falsification. When it happens that the scythe cuts an unfortunate theory down to the roots, the reaction of the majority of scientists is to stress the weakness that made its fate predictable and to rejoice in the expectation of a better theory to come.

The strong difference in emphasis on this matter between a number of philosophers and most scientists cannot be understood if one believes in Kuhn's ideas of paradigms and scientific revolutions. They describe in fact an outdated stage of science and it has been a long time since the last revolution took place in physics. *Consistency* describes contemporary science much better than the unrest of paradigms. During a large part of the twentieth century, the story of physics has been dominated by a search for consistency and its recognition, and our reflections must certainly take that into account.

## Some Examples

Much of the history of scientific controversies was dominated by questions of logical or mathematical consistency, but a few examples are particularly clean. In Maxwell's equations, for instance, the most important addition to the laws that were already known was the "displacement current," which is the time derivative of the electric field that must be added to the current density in the calculation of the magnetic field. Maxwell had guessed its existence through a careful investigation of an intricate model, but it became universally accepted when it became clear that it ensures the consistency of the evolution equations with the conservation of charge. It led immediately, by the way, to the prediction of electromagnetic waves.

147

The theory of special relativity came entirely from considerations of consistency, and as a matter of fact in two different ways. Einstein stressed the lack of consistency between the invariant velocity of light resulting from Michelson's experiments and the rules for combining velocities when time and space are absolute. Poincaré considered the group of invariance of Maxwell's equations (now called the Poincaré group and containing the Lorentz group as a subgroup). He compared it with the Galilean group of invariance of nonrelativistic mechanics and noticed their incompatibility. Einstein's relativistic theory of gravitation originated also from a lack of consistency, since Newton's instantaneous gravitational force disagreed with special relativity: violations of causality could appear after a simple change of reference frame.

The search for consistency between special relativity and quantum mechanics again had a central role in the development of particle physics. It began with Dirac's equation in 1928, which generalized the nonrelativistic Schrödinger equation to the case of a relativistic electron. Since the Schrödinger equation is linear in the differential operator $\partial/\partial t$, relativistic invariance requires linearity in the space derivatives $\partial/\partial x$, and some algebra implied that it meant writing the energy $E = \sqrt{p^2 + m^2}$ as a first-order polynomial in the momentum $p$. When space is one dimensional, so that $p$ is a number rather than a three-vector, the coefficients $a$ and $b$ of this polynomial $ap + bm$ must be matrices such that $\alpha^2 = 1$, $\beta^2 = 1$, $\alpha\beta + \beta\alpha = 0$. In the case of four-dimensional space-time, the matrices had to be four dimensional, the resulting equation automatically describing a spin-1/2 electron and considerably improving the understanding of the hydrogen atom. This beautiful result was obtained when the mathematical framework of quantum mechanics was not yet definitely fixed, and its elegant juggling between wave functions and matrices seemed marvelous.

The next steps in the story of Dirac's equation are also interesting, because they make plain a significant difference in the meaning of consistency, when considered from the standpoint of physics or of mathematics. One might say that physics wants consistency, but it is ready to wait a bit longer for logical steadiness when some crazy idea agrees with experiment. Mathematics, on the contrary, does not even exist when consistency is lacking. The Dirac equation had a serious drawback: one-half of its solutions corresponded to negative energies! Dirac's proposal for getting out of the rut looked absolutely crazy: he assumed that all the negative energy states are occupied in ordinary vacuum and, hey presto! He predicted that an antielectron should exist. And it exists! Anderson found it in cosmic rays and logic could wait a bit longer.

The slow search for consistency continued with many ups and downs. There were beautiful improvements such as Wigner's mathematical synthesis of special relativity and quantum mechanics through the Poincaré group: he got rid of negative energies and found that consistency implies the existence of mass and spin. There was also quantum field theory, which explained the success of Dirac's equation because it governs the electron *field* and not a wave function. But quantum field theory stumbled against infinities, which popped out everywhere. The way out of these infinities, renormalization, is another wonderfully crazy piece of quasiempirical mathematics, self-consistent though showing no relation yet to the axiomatic part of analysis (it becomes a bit more wonderful and slightly less crazy as time goes on, but its foundation always remains around the next corner). The removal of infinities through renormalization works perfectly well for some interactions, and only those interactions, so that it could provide valuable guidance during the construction of the standard model of quarks and leptons.[5]

[5] See, for instance, Hoddeson et al. (1997).

## *Present Tendencies*

It would certainly be interesting to recount the history of particle physics until the discovery of the standard model as a double quest, experimental and theoretical; it was experimental during a long search for hints and when essential verification tests were at stake, theoretical also and even mathematical when the search for consistency led the way. The convergence of physical and mathematical ideas would be particularly striking. Long after Dirac bypassed Hilbert spaces and invented the delta function—before Laurent Schwarz's distribution theory—one would see the theory of group representations (simultaneously developed under the impetus of mathematicians and physicists) and fiber spaces (originating in geometry and in the gauge theories of the standard model). The examples would be many, and the people who are familiar with them tend to forget the difference between the study of mathematics and of the laws of physics.

The main present problem in fundamental physics is again a question of consistency, which now involves quantum theory and general relativity. It was more or less left out of consideration until very recently, because the corresponding effects were expected to take place at very small scales in space and time, very far from anything accessible to the present experimental possibilities. The length scale, for instance, is the Planck length $\sqrt{G\hbar/c^3}$, which is of the order of $10^{-33}$ cm and results from a combination of the three fundamental constants $G$, $c$, and $\hbar$ (i.e., the gravitational constant, light velocity, and Planck constant). A few tantalizing problems serve as guides for this new research. According to general relativity, the curvature of space-time is generated by its energy content, but the particles and fields carrying the energy obey quantum rules. There is therefore an inconsistency in assuming space-time to behave according to classical reality. A simple idea, which has been much developed, was to "quantize" general relativity. It meant finding a quantum theory that would coincide with general relativity in the limit where Planck's constant is considered very small. General relativity deals with the metrics of space-time, which is a field (since it consists of quantities depending on four space-time coordinates), and its quantized version is accordingly a quantum field theory, which shares many common features with the gauge theories of the standard model. It does not share, however, their nicest feature, which is the possibility of getting rid of meaningless infinities through renormalization. It is therefore, unfortunately, an inconsistent theory.

One may remember that one of the main guides in the discovery of the standard model of quarks and leptons was precisely the possibility of renormalization. The search for a renormalized quantum form of general relativity has therefore become a strong incentive for a new quest. Tantalizing results were obtained in that direction by string theory, which we can only summarize too briefly (see Greene 1997): There are no pointlike particles, but matter and radiation are made of microscopic strings (or objects of a higher dimension); their quantum theory can be consistent only when space-time has ten dimensions, so that six dimensions are supposed to curl up at the scale of Planck's length, leaving as observable a four-dimensional space-time. Finally, lo and behold! Out of this fairy tale comes a renormalized general relativity.

Noncommutative geometry is probably a more systematic approach, which has been particularly pioneered by Alain Connes (1994, 2000). It is perhaps more in Einstein's spirit, who never jumped at an answer or guessed it, but let a question generate its own answer through pure consistency. Anyway, whatever the approach to be followed, one is contemplating the first attempts in the whole history of science at letting a question come to the forefront of research, without even the perspective of some empirical backing. Such an enterprise would have been considered metaphysical not so long ago. The fact that brilliant and dedicated people work actively on it, and encounter promising physical and mathematical discoveries on

151

their way, indicates that a new philosophy of physics and mathematics is growing up after the recent lessons of the standard model. When explicitly formulated, this philosophy will probably appear as a variant of physism, either in the form that was proposed by Connes himself (see Changeux and Connes 1995) or in the version of the present book.

## The Laws, Timeless and Creative

### Timeless Laws

The fundamental laws of nature have always been considered immutable, for reasons apparently closer to human nature than cogency. There were long ago specific arguments relying on the perfection of God's work or more directly on ontology, as in Aristotle's principles or Leibniz's constructions. Many thinkers, including Descartes and Hume, associated the laws with mechanisms, which should work always and everywhere exactly in the same way. Astronomy and astrophysics provided the first objective reasons for assuming universal, and therefore timeless, laws, when Newton discovered that the same dynamical laws govern every motion on the Earth and among heavenly bodies, or when Kirchhoff extended the laws of optical absorption and emission to the sun and the stars.

The existence of astrophysics, and particularly its consistency with observational data, is by now the most cogent argument for the timeless character of the laws. Dirac noticed, however, that some "constants" entering into the expression of the laws (such as the gravitational constant or the fine-structure constant) could evolve with time while the form of the laws would not change. But no data have substantiated this idea until now, though some astrophysicists continue to keep it in mind.

By and large, this question was associated with a more general one, which asks what bounds could limit the validity of the laws that we know. Since general relativity, for instance, can be applied to the totality of space-time, the only possible check of its limits consists in taking its consequences as far as possible, until one might get a contradiction with observation. No such divergence has been found up to now; on the contrary, there is a remarkable agreement between astronomical observations and the conjunction of the two "standard" models describing, respectively, the big bang and the quantum particles.

These results, by the way, put philosophy into a rather awkward position. Its role in the critique of reason should incline it to warn us against excessive enthusiasm and to stress the possibilities of error in the conclusions of science. But, on the other hand, the two-century-old tendency of philosophy to condemn any form of metaphysics is difficult to maintain in view of these extraordinary observations. Perhaps the best compromise would be to accept the universal character of the laws as a highly plausible assumption and to work out its consequences. That will anyway be the position of this book.

### Creative Laws

A fascinating aspect of the timeless character of the laws is their predictive content, which made them keep in reserve for a long time the creative power of their potentialities. This feature is often illustrated by the fact that the potentiality of life was contained in the laws long before life appeared; it is sometimes interpreted as a form of finalism, and one often speaks of emergence or creation when such an inner potentiality blossoms. A few examples will show what it means better than any long comment. If one considers, for instance, the standard model of quarks and leptons, it nicely explains everything in nature around us with its nuclei, atoms, and molecules,

153

but there was obviously an era at the beginning of the universe when all these features were still pure potentialities. No atom existed when the universe was not yet half a million years old. It would be easy to spell out a long series of beginnings waiting in the same way inside the bud of laws: the beginning of matter from inside radiation, its concentration through gravitational attraction, giving rise to galaxies, stars, and planets, the beginning of heavier nuclei in stars, and, of course, the beginning of life.

One of the most fascinating beginnings was probably the origin of matter, although it is yet somewhat speculative. The point is that three families of quarks and leptons have been discovered, and nobody knows why from the standpoint of consistency. Apparently, one family would have been enough to explain everything we see presently in nature, and it seems difficult to resist asking the question "Why three?" It turns out that, if there had been only one or two families, the consistency constraints on the standard model of particles would have forbidden the existence of an interaction violating $CP$ (the product of the operation of charge conjugation $C$, which exchanges particles with antiparticles, and of the parity operation $P$, which reverses the direction of the space axes). On the other hand, three families permit a violation of $CP$. But that implies that there was a period, near the origin of the universe, when more quarks than antiquarks were produced and ultimately, much later, that excess of quarks over antiquarks resulted in the existence of matter in the universe. In other words, matter and antimatter would have annihilated each other long ago if there had not been three families of quarks. It seems accordingly that the existence of matter, which means everything we see including the sun, the earth, and every live being, was in some sense predictable from the number of quark families. Further investigations will probably decide whether this idea is valid or not, but if taken as a paradigm, it shows plainly the predictive or creative character of the laws, which is sometimes hidden in obscure detail.

## *The Laws versus Time**

The relations of the laws with time are deep and subtle (Hawking 1988; Penrose 1997). The laws speak of time, since they describe it. The initial singularity of space-time in the big bang models of the universe results from general relativity and it is consistent with a beginning of time (something unheard of previously, except in St. Augustine's *Confessions*). Quantum gravity suggests, on the other hand, that time, as we understand it classically, did not begin immediately but came out after an initial period of quantum space-time foam (but what is meant then by "after"?). There is a definite cosmological direction of time, because the initial state of the universe had a high degree of order, which later could only decrease; there is also a time direction in thermodynamics, which is closely associated with decoherence. There is nonetheless no direction of time in the quantum laws themselves and, since Newton, time has been considered in physics as a simple, perhaps too simple, parameter.

We just saw the potentialities of the universal laws of physics and how they are disclosed one by one in the course of time. We might also have used the language of Henri Bergson and spoken of the creativity of time rather than of the laws; or we might have tried to illustrate Martin Heidegger's philosophy as a kind of complementarity between Time and Being.[6] It would have been an attempt at reaching something essential in time, of which no law can take hold completely, but it would have probably been a gratuitous game. The existence of time anyway spells out the only obvious difference between a law of physics and a pure mathematical statement, since time actually develops the potentialities of the law, whereas only a logical construction can develop those of the corresponding mathematical proposition.

---

[6] Being fades away when disclosed in an instant. (Heidegger).

## THE LAWS: INACTIVE MATHEMATICAL FORMS

### *The Fundamental Laws Are Inactive*

The Newtonian laws of physics were causal and they expressed an effect of action. Forces acted, inertia resisted, and motion resulted from their antagonism. Classical reality has deeply imprinted this idea of a universal action in the human mind, and one may remember how, in Aristotle's physics, things did not move but were moved under the impulse of the prime mover. During the Middle Ages, angels were in charge of pushing the arrows along their courses (what poor angels, and poor guys who happened to stand on a trajectory). Modern biology continues to think of the laws as acting through mechanisms (and there is of course no objection to that view since it is true of everything macroscopic, beyond quantum decoherence). There is again something reminiscent of action in general relativity: when a falling object follows a space-time geodesic, its motion is determined by the curvature of space-time, and this curvature is produced by matter (or, more precisely, by the energy-momentum of matter and radiation). Finally, it can be said that matter and radiation act.

The situation is quite different in quantum mechanics, although it did not become clear till the early 1950s. One considered earlier in atomic physics that a charged atomic nucleus creates a Coulomb field, which acts on the electrons. The electromagnetic potential vector was even considered with diffidence for a long time, because it does not act by itself and only the electric and magnetic fields do act. The discovery, by Yakir Aharonov and David Bohm in 1961, of the effect of the potential vector on electron interferences in a region where the magnetic field vanishes was therefore a surprise for people who still thought in terms of action. One may also recall the mental representations of virtual processes, which were discussed in chapter 8. They were another emergence of the idea of action, since these processes "looked as if" something was acting at an inaccessible level.

The fact that there is no action in quantum mechanics became clear with Feynman histories, which are the ultimate form of quantum field theory. It became still more obvious with the later development of the standard model. In Feynman's framework, there are histories, they interfere, and the observed data result; nothing acts.[7] Action and causality appear only beyond decoherence, which is an interference effect. This lack of action is more easily understood from a simple argument, which does not enter into the structure of the quantum laws and consists in looking at an isolated hydrogen atom. What a poor thing this atom is: just two particles, one of them made of three quarks and the other a charged point. Through what action, what mechanism, could it achieve the production of the minutely precise and reproducible effects that can be measured on it, and which are predicted by the laws?

Maybe further research into the mysteries of space-time at the scale of Planck's length will change the situation, but probably not by restoring action at the lowest possible scale. A cautious philosophical approach, anyway, can rely on the laws only as they are presently known and, from that standpoint, *the fundamental laws of nature are pure mathematical forms accounting for the phenomena though providing no cause for them and showing no action*. This statement is worth special emphasis, because it provides the most cogent argument for identifying the fundamental laws with mathematical forms, i.e. with mathematical objects.

---

[7] A trivial caveat is the occurrence of a mathematical quantity that is called "action" in the phase of a Feynman history; but it has nothing to do with the fact of acting; it is a remembrance of a classical quantity that was called "action" in the "principle of least action" and its name is only a by-product of the history of physics.

CHAPTER TWELVE

## *The Invariant Form of the Laws*

The fact that quantum laws are pure forms is reflected by the invariance of their expression, independently of the level at which they apply. They have the same form when they deal with quarks or with atoms, and this permanence of form, established in the case of atoms in 1926, has withstood an extraordinary extension during three-quarters of a century. When one compares, for instance, the energy scale of the particle collisions that are described by the standard model with the most precise details of the hydrogen spectrum (the "Lamb shift" with its ten measured and computed figures), one finds a ratio of $10^{27}$! Twenty-seven orders of magnitude over which the quantum laws govern with identical rules; this is about the ratio between the radius of the universe and the size of a person.

When speaking, however, of an invariant form of the laws, one does not mean that there is a unique way of writing them. We have seen, on the contrary, that one can rely equally well on the Schrödinger equation in a Hilbert space or on Feynman histories. This means that there exist different expressions of the laws, but they are *mathematically* equivalent, so that they constitute a unique mathematical object. The Form does not change, whatever the system it describes.

The significance of this invariance is best understood once again by looking at Feynman histories. We saw in a previous chapter that the amplitude of a physical process is the result of a sum over histories, but we did not mention that the content of the histories is extremely flexible. They can describe equally well particles or quantum fields, at various levels. When considering an atom for instance, there is no objective criterion for choosing the content of histories, which remains a matter of convenience. One might use the basic fields of the standard model, which involve the fields describing quarks, gluons, leptons, and intermediate bosons (including photons). One might also consider a simpler description involving a nonrelativistic nucleus together with the fields of photons and electrons, as in

158

quantum electrodynamics. The first description is in some sense more fundamental but also much more burdensome, while the second one is certainly more convenient when one is interested in the details of the spectrum of the hydrogen atom. When considering chemical binding, on the other hand, it will be even more convenient to use a still simpler description of atoms with no quantum field and only nonrelativistic particles. In all three cases, the formalism of histories is exactly the same.

The mathematical relation between these various descriptions relies on the fact that one is *summing* over the Feynman histories. This sum can be performed over the various positions of a mathematical particle (a clone), as time goes on. Or it may be a summation over the values that a field can take at every point of space-time. Fundamentally, one would expect that the most exhaustive description should rely on all the existing fields at every point of space-time, but we are not even sure of knowing what these fields are. Surely, if string theory is correct, string fields are also relevant. But some fields can be irrelevant for a definite process: quarks are irrelevant for a study of the hydrogen spectrum, and electron fields with their virtual effects are irrelevant in the case of chemical binding. The beauty of an invariant formalism is that one can "integrate out" the irrelevant fields, i.e., get rid of them at the price of an approximation, the result being still a Feynman sum over histories with a fewer number of field variables! The *form* of quantum mechanics remains unchanged under this elimination.[8]

---

[8] To avoid any misunderstanding, one should add that the invariance of the quantum form under the elimination of irrelevant degrees of freedom is not obvious. The error of considering it obvious is partly responsible for the long time it took to discover decoherence. One should justify in every special case the irrelevance of suppressing some degrees of freedom, and it does not apply when one suppresses those of the environment. On the contrary, decoherence results from that suppression, and from there on the form invariance of the laws disappears. This point is further discussed in the next chapter.

The form invariance of the quantum laws can be envisioned from two opposite directions in the following sense: One can integrate out irrelevant degrees of freedom, as we just explained, or one can insert new degrees of freedom when they are required. The relativistic electron field was added, for instance, to the description of atoms when the Lamb shift became a valuable subject of research.[9] Previous results in atomic physics nonetheless remained perfectly valid since one could retrieve them by "integrating out" the electron-positron field. With hindsight, the historical development of quantum physics may even appear as a series of insertions: the quantum fields of electrodynamics were first inserted into atomic physics, then came those of the electroweak interactions, and a little bit later the quark and gluon fields. The story of insertions is probably not ended since the success of the standard model is often interpreted as a compatibility with still unknown deeper layers of fields, which are not yet manifest at presently attainable energies. A more complete discussion would add that the gauge theories of the standard model have much improved our knowledge of the mathematical significance of the laws, particularly about symmetries. New and deeper questions emerged from this progress also, but this other aspect of the known and unknown faces of reality would take us too far into highly evolved theories, and they are not essential for our purpose.

## The Laws Deal with Pure Potentialities

The fundamental laws deal with potentialities, or possibilities, and not in any case with actuality. This statement again looks

[9] The Lamb shift, which was discovered and measured by Willis Lamb after World War II, is a difference in the energy levels of two states of the hydrogen atom, which were predicted to coincide according to the Dirac equation. This observation was one of the starting points of quantum electrodynamics.

to be at variance with classical tradition. There was no room for potentialities, options, chance, or odds in the physical world at the apex of classical physics, when determinism was elevated to the rank of a principle. Every process was the unerring unrolling of the consequences of previous facts. Laplace, who proclaimed determinism and also worked on the foundations of probability theory, was particularly clear on this matter: the motion of a die, for instance, is completely determined by the initial throw, the velocity and the rotation of the die when it leaves the player's hand or the dice box. It turns out, however, still according to Laplace, that the finest details of the throw are inaccessible to our control or our observation, and we do not know them well enough to apply determinism in practice. The notion of chance and odds amounted therefore from this standpoint to the recognition of human ignorance and limited abilities.

Max Born discovered the probabilistic meaning of the wave function when considering the collision of an electron with an atom and he concluded his paper with some cautious words on the loss of causality:

> Here the whole question of determinism comes up. From the standpoint of quantum mechanics there is no quantity which in any individual case causally fixes the consequences of the collision; but also experimentally we have so far no reason to believe that there are some inner properties of the atom which condition a definite outcome of the collision. Ought we hope later to discover such properties (like phases or the internal atomic motions) and determine them in individual cases? Or ought we to believe that the agreement of theory and experiment—as to the impossibility of prescribing conditions for a causal evolution—is a preestablished harmony founded on the nonexistence of such conditions? I myself am inclined to give up determinism in the world of atoms. But that is a philosophical question for which physical arguments alone are not decisive.

To speak of "a preestablished harmony" has a smell of Leibnizian metaphysics, but it expresses magnificently the relation of the laws with reality when one thinks of their lack of action and their dealing with purely potential events. Even so, the relation of the quantum laws to potentiality is in fact more general than noncausality. The Feynman histories shed vivid light on that question and they have shown that potentialities go much deeper than the random occurrence of an event at the end of a measurement. Each history represents a time-ordered sequence of virtual events rather than only a final one, and they are perfectly arbitrary; they know absolutely no constraint and no determination. They are still more out of control than the wave functions—which were interpreted by Born as probability amplitudes—since the time evolution of a wave function is *determined* by the Schrödinger equation. The Feynman histories express total freedom, and also something more, which is a harmony born of freedom. Total freedom does not mean, on the other hand, a lottery where a superdie would be thrown with a history painted on every face: the die would finally show a unique history and the result would not be preestablished harmony, but a pandemonium where anything could happen without any rule. No! The histories are not mutually exclusive events, but mathematical forms. They add together their vibrating amplitudes in a perfect harmony, from which order can arise through variety.

John Wheeler often asked the provocative question: "Why the quantum?" Perhaps an answer would be that the quantum laws are able to unify the potentialities of total freedom with the existence of a high type of order. Presumably, no other law could do that.

## Conclusion

Our brief examination of the character of the laws of nature— at least as they are known presently—shows perhaps more clearly their depth and strangeness than a more explicit and more scientific study of the laws themselves would have done. They may leave us with the impression that we do not understand what kind of reality stands behind them, but that should be no surprise if one remembers some relevant comments by Wittgenstein (see Bouveresse 1988): reality is primary and it cannot be understood from outside. One may add now that our mind stands within classical reality, and the fact that reality, or some reality, or the nucleus of reality, has become different means that we cannot understand it fully (in the Latin sense of understanding, *comprehendere*, or seizing something in its totality). We noticed along the way that time is an intrinsic, nonformal, component of reality (reminding us of Bergson and Heidegger), but it is clear that philosophy is still far from coming to terms with the full extent of reality and there is presently no satisfactory entry that way.

*The laws expressing the regularities of reality are much more accessible to understanding than reality itself.* They exist and they are consistent, they are pure relations, apparently inactive, and allowing every possibility on an equal footing, and that is practically the definition of mathematics, or at least a part of it. Time spells out the hidden content of the laws, just as a mathematical analysis develops the consequences of a system of axioms. The laws appear as pure forms; they are forms belonging almost entirely to the realm of mathematics. They are prior to mathematics, however, just as reality is absolutely prior to anything. *The idea that the laws of nature encompass mathematics therefore appears finally as simple and attractive.* This statement is the main conclusion of the present chapter.

# The Character of Classical Reality

Everything around us is macroscopic, including what can be seen with optical or electronic microscopes. This is the domain of classical reality, with the characteristics we met in chapter 2: uniqueness, a definite direction of time with a well-defined past, localization in space, continuity, separate objects, and finally causality. All these features were embodied in the classical laws of science, but they were of course more basic and already constitutive of common sense. They show a sharp contrast, however, with the characteristics of the fundamental laws in the previous chapter. This opposition was a bewildering source of perplexity for a long time and it began to be understood only recently, with the discovery of the decoherence effect. This effect, which we discussed in chapter 10, produces such a drastic change in the character of the laws that one might speak of a transmutation, or almost a change of nature.

Although several significant questions are not yet completely clarified, the transmutation of the laws is important for physism. We already encountered some of its long-ranging effects on several occasions. The opposition between sophisticated mathematical objects—such as nondifferentiable curves or many-dimensional spaces—and the supposedly simple features of nature was instrumental in the separation of mathematics from early versions of physism. When it was later recognized that very abstract mathematics were necessary for the formulation and utilization of the fundamental laws, it could have been a significant clue, but the difficulties in the interpretation of quantum mechanics precluded a philosophical extension. This is why I believe it worthwhile to come back again briefly to this question, even though it was already introduced for its own sake in part 2.

Another reason for stressing that point is to dispute a too frequent opinion about the relation between mathematics and the science of nature, which is sometimes expressed by very competent persons and of which there is no better example than the following quotation from Ruben Hersh (1979): "We may ask how these [mathematical] objects, which are our own creations, so often turn out to be useful in describing aspects of nature. To answer this specifically in detail is important and complicated. It is one of the major tasks for the history of mathematics, and for a psychology of mathematical cognition. . . . The answer, however, is easy and obvious. Human beings live in the world and all their ideas ultimately come from the world in which they live—refracted through their culture and history, which are in turn, of course, ultimately rooted in man's biological nature and his physical surroundings. Our mathematical ideas fit the world for the same reason that our lungs are suited to the atmosphere of this planet."

I chose this quotation because I respect its author, and this respect makes it easier for me to say that I understand why a learned mathematician might hold this opinion. But it shows also that one must explain why this apparently attractive point of view should now be considered as too naive and obsolete. The present chapter will therefore be devoted to a few additional comments on classical reality, mainly to stress how nontrivial it is. I will also take this occasion to mention some present limitations in our understanding of it. Both comments and limitations would be, of course, more thoroughly worked out in a book on physics, but I will try to stay within the boundaries of the present project.[1]

## Conceptual Constructions as an Inverse Process of Abstraction

It may be convenient to begin with a few philosophical considerations. It is generally agreed that understanding classical physics was one of the difficult tasks facing quantum mechanics. Although this assertion might look paradoxical at first sight, a little reflection shows where the difficulty lies. The discovery of quantum mechanics was the outcome of a long process of abstraction, starting from classical knowledge. It involved a replacement of physical quantities by noncommuting operators, a state vector in Hilbert space or a wave function was substituted for a classical state involving both positions and momenta, causal dynamics was replaced by a Schrödinger equation or interfering Feynman histories. There was also an immersion into deeper layers of reality, in which one went from the usual objects to the atoms and their constitutive particles, then to more quantum fields, before arriving at quarks and gauge theories. That was a long way on, but the way back to the classical world was not much easier. We mentioned already the fundamental importance of decoherence, but the recovery of classical physics from the fundamental quantum laws requires almost as much abstraction as before. For instance, the return from noncommuting observables to ordinary classical variables is a subtle mathematical problem relying on microlocal analysis, which is a rather recent addition to the mathematical corpus. Decoherence theory, which explains the

[1] For more details, see Griffiths (2002), Giulini et al. (1996), and Omnès (1999b).

transition from a quantum state to a classical one, is still in its infancy. We have models, phenomenological theories in good agreement with experiment, but not yet a direct proof that decoherence is a mechanism of coherence loss, as we strongly suspect. We do not know yet how to identify fundamentally, in a quantum Hamiltonian, the observables that will behave classically at the end. We know what they are, of course, in any realistic case—because we "see" the outcome—but we have no explicit algorithm for picking them up at a fundamental quantum level. In other words, this part of quantum mechanics is reasonably well understood from a practical standpoint, but the necessary mathematical tools are not yet available for making it into a fully consistent construction.

It seems that one must go back over every step taken during the quantum deconstruction of classical reality, in order to reconstruct it on the basis of quantum laws. Of course, I am aware that every technician says something similar about the difficulties of his job, but there is something special in this one and its relation to abstraction. The point is that physicists proceeded by means of abstraction when they built up quantum theory, and now they must again abstract some aspects of this abstract quantum stuff when they reconstruct the main features of classical reality. This is in some sense a hyperabstract procedure and it would be just a game for the fun of hyperspecialists if its main goal were not to interpret what is most obvious and most intuitive for the human mind: I mean plain, ordinary reality.

This apparently circular approach, from conspicuous reality to hidden quantum laws and back to conspicuousness, is probably interesting for the philosophers who try to understand abstraction. It shows clearly, at least, how mathematics is involved in an essential way when one attempts to reach a consistent understanding of nature. By the way, nothing could differ more than that from Hersh's matter-of-fact vision of the relation between mathematics and reality.

## Determinism and Causality*

### *Determinism*

For many years, one could read learned discussions in philosophy books of a supposedly total opposition between classical determinism and the quantum probabilistic approach to the world of particles. This question was indeed very troublesome and Einstein's famous assertion, "God does not play dice," was not meant as a joke. The new approach through decoherence, microlocal analysis, and consistent histories is particularly interesting because it shows how a bit of mathematical sophistication can sometimes bring a verbal contradiction to an end.

One may consider as an example the motion of an electron in a television tube. Electric and magnetic fields guide its motion and the system looks perfectly deterministic, even though the electron is a quantum particle. There are, however, sound reasons to believe that the quantum laws are fundamental, although not in any way deterministic, and this apparent contradiction is worth understanding.

We may try first to make the question cleaner. Surely, we will not assume like nineteenth century people that the electron is a moving space point with a precise position and a sharp velocity. To explain what is observed, and taking into account the conditions of observation, it should be enough to understand why the electron reaches the screen at a somewhat well-defined point with a somewhat well-defined velocity. Let us therefore consider that, according to quantum laws, its initial position is given with an error $\Delta x$ and its initial momentum with an error $\Delta p$, the product $\Delta x \cdot \Delta p$ being large with respect to the Planck constant $h$. An analogous description will be assumed for the end result when the electron hits the screen. The main question then becomes: Is there a deterministic relation between these initial and final properties?

Some standard results in microlocal analysis can be used to show that definite projection operators $P$ and $P'$ (in the Hilbert space of the electron wave functions) can be associated with these properties, so that one can rely on the formalism of Griffiths' histories as met in chapter 11. Since we consider quantum dynamics to be fundamental, we must derive classical dynamics from it. This problem is well known in microlocal analysis and the answer is given by a theorem which was first discovered by Yuri Egorov. The relation of quantum mechanics with determinism then boils down to the question: Is the quantum projection operator $P$ expressing the initial condition transformed through unitary quantum dynamics into the projection operator $P'$ expressing the final property, as predicted by classical determinism? The answer, using Egorov's theorem is "yes," or more precisely—according to the logical framework of consistent histories—the initial classical data imply the final ones, and conversely.

This result looks exactly like classical determinism since it means that the initial and final properties are logically equivalent (implying each other): given the initial data, the final ones follow; conversely, one can infer the initial data from an examination of the final ones. There is a subtle difference, however. The exact consequence of Egorov's theorem is not, for instance, that the initial data exactly imply the final ones, but that this inference has only a very small probability of error. In the case of an apple falling from a tree, the exact result is that the apple will fall under the tree, except for an extremely small quantum probability for observing a different outcome where the apple would fall elsewhere, begin gravitating round the earth, or any other crazy behavior. *The final answer is that determinism is valid with a very small probability of error.* The quantitative evaluation of the error depends on the case at hand and it is usually very, very small. The longstanding verbal opposition between classical determinism and quantum randomness is therefore dissipated when determinism is expressed in a probabilistic way: it holds in most practical cases with such a small probability of error that it can be confidently considered as valid.[2]

## Causality

The example of an electron in a television tube dealt with a microscopic system undergoing no friction. The result was not yet causality, since this concept applies more properly to a macroscopic system and involves a specific direction of time according to which a cause is prior to an effect. This question was discussed at length by Omnès (1999b) and is more difficult than nonfriction determinism. Roughly speaking, one may say that decoherence is an irreversible process in which phase information is continuously lost into the environment. An important consequence of this irreversibility is the occurrence of a definite direction of time in the framework of consistent histories (it turns out that most histories are consistent because of decoherence when the events follow each other in one time direction and they are not consistent when the time order is reversed—or the film runs backward, so to say). The consistency conditions imply that the directions of time according to thermodynamics and to decoherence must run in the same direction as decoherence. Causality is then a consequence of the validity of classical physics following decoherence.

It should be mentioned that these results have not yet been established with the full care and rigor that their importance would warrant, but it might also be remembered that the investigation of decoherence and its many important outcomes is still in progress.

---

[2] The error resulting from the assertion of determinism increases when one tries to make it more stringent. The probability of error becomes unbearable at atomic scales, for instance. There are also limitations on the validity of determinism in the case of chaotic systems: classical dynamics is not reliable when chaos approaches the scale of Planck's constant.

## ON THE UNIQUENESS OF CLASSICAL REALITY

The question of sharpness—or uniqueness—of classical reality is particularly significant when a quantum measurement occurs. A "measurement" is meant in the present case as a natural or manmade phenomenon in which a quantum superposition is amplified to a macroscopic scale and almost simultaneously broken through decoherence.

When no such phenomenon occurs, the physical conditions are plainly those of classical reality, whose uniqueness raises no serious problem. Its derivation is nonetheless technical, like everything resulting from quantum mechanics. It relies on the fact that the logical framework of consistent histories is unique when one is dealing with macroscopic properties (the ambiguities arising from quantum complementarity do not occur in that case). In other words, the classically meaningful histories (involving only classical properties) build up a unique framework describing a sharp reality, and the logic of consistent histories coincides with the common sense logic of classical physics.[3]

### The Case of Quantum Measurements

When a quantum measurement or an equivalent effect occurs, the theory of decoherence shows that the various possible results of the measurement must obey standard probability calculus, with no trace of quantum superpositions (no dead-and-alive Schrödinger's cat). The fundamental quantum laws do not predict, however, how a definite unique result will come out from a measurement. This question is still a matter of controversy although not directly related to the topics of the present book. I will therefore only mention the

---

[3] The fascinating relation of common sense to quantum foundations has been discussed in some detail by Omnès (1999a).

competing theses together with the names of some of their main proponents.

1. Quantum mechanics is intrinsically a probabilistic theory and it cannot therefore account for the result of an individual trial (Gell-Mann, Hartle, Griffiths).
2. In any case, a definite measurement result becomes logically separated once and for all from the alternative ones because of consistency. A mechanism that would determine the issue of a measurement would therefore be unobservable (Griffiths).
3. Physical reality is not unique. There exist many noncommunicating universes in which the quantum alternatives proceed to all their possible consequences (Everett).
4. A random effect producing a spontaneous reduction selects a definite result (Ghirardi, Rimini, Weber).[4]
5. Quantum mechanics as it stands now is not a complete theory and it does not encompass general relativity. A specific reduction process enforces the uniqueness of space-time and it must involve in some way the gravitational constant (Penrose, Pearle, Karolihazy, Frenkel, Diosi).
6. Bohm's proposal, involving the existence of real particles with definite positions, will be found finally to be consistent with special relativity and field theory. It will then ensure automatically the uniqueness of reality (Dürr, Goldstein, Zanghi).

When one tries to extend the concept of physical reality to the quantum domain, many alternative views result from these various proposals. The question of complementarity makes the problem still more complex, and it would be foolish to assert by now what can be meant by the word "reality," in a more or less universal sense. One can at best propose speculations or resort

---

[4] The effects one assumes in cases 4 and 5 cannot, however, be quicker than decoherence and they come after selecting a result, with no observable consequence as observed in answer number 2 (see Griffiths 2002).

to some kind of renunciation. I consider it fortunate in these conditions that an answer to this question is not necessary when one tries to deal with the philosophy of mathematics. One should stress once again that no empirical data have been found up to now to disagree with quantum mechanics, so that a purely mathematical vision of physics relying only on its basic laws is perfectly consistent. As a matter of fact, it is enough from that standpoint to understand the emergence of the main characters of classical reality from the basic quantum laws, and this precise understanding has become much better in the last few decades. Conversely, this understanding was necessary before one could revive physism as a reasonable approach to the foundations of mathematics, which explains the origin of the present book (I dare say I expect the question of physical realism to be easier to handle within the framework of physism, but the opposite approach is not yet dependable).

# PART FOUR

## PHYSISM

**E**verything is now ready to draw the conclusions of this book. We shall do it in five chapters. The first one will be only a reminder of the main previous theses on the nature of mathematics: those of logicism, axiomatism and formalism, intuitionism, Platonism, and quasiempiricism, with a few remarks on the standpoints of information theory and cognition sciences. This chapter is meant only to set the stage. The second one asserts the main thesis, namely, mathematics belongs fundamentally to the laws of nature. This is called physism, remembering that the word is patterned on "logicism" and its root is understood as *physis*, the Greek word for nature. This proposal does not mean, of course, that a mathematician is investigating the natural laws when he or she works, but that the underlying reason behind the mysteries of mathematics is due to its direct, consubstantial, relation to the laws. These mysteries—why is mathematics so vast, consistent, why it is so fecund, and yet remains an objective entity rather than a flexible work of art—these mysteries, I say, reflect faithfully identical, or almost identical, properties of the laws, which themselves stand as the unique source of meaning.

The third chapter will compare the proposed thesis with the existing ones. The fourth chapter will be a discussion of the thesis and finally the fifth one will venture into the slippery domain of ontology. It will be short and cautious, because I tried as much as I could in this book to look only at the facts and analyze them; ontology, per se, is therefore foreign to it, but a few words can do no harm.

# The Philosophy of Mathematics

The philosophy of mathematics is a wide and deep subject with a long and rich history. There are excellent books on this topic and there would be no point in our trying to copy them (Benacerraf and Putnam 1983; Van Heijenoort 1967). Some main trends only will be described in this chapter in a cursory way to provide a few benchmarks for later comparison with physism.

## Logicism

### *The Rise and Fall of Logicism*

Toward the end of the nineteenth century, mathematicians were embarrassed by a superfluity of good things. They were trying to master non-Euclidean geometries. Geometry itself had to be redefined, after much mingling with algebra and analysis. Topology took unprecedented turns in many directions. Infinity had to be mastered and clarified. Set theory seemed ready to encompass the whole of mathematics, but although generality and abstraction were pushed forward everywhere, they counteracted intuition and doubts were raised about their validity. Logic was showing signs of weakness in many places and it had to be placed on firmer bases to face the challenge. Then Frege came.

He was the main architect of logic. There would be no point in recalling here the foundations of mathematical logic as he straightened them out, since most of them are now taught in mathematics or philosophy courses. It will be enough to mention that he introduced the truth-functional proposition calculus and the universal and existential quantifiers ($\forall$, $\exists$) and founded the notions of variable and function. He defined inference as the direct action of a set of logical rules on the form of propositions and it was clear after him that mathematics is a pure science of relations with no privileged content.

Frege also had strong views on the philosophy of mathematics, and he propounded logicism, a thesis according to which logic is the sole foundation of mathematics. The key assumption was that every property defines a set: namely, the set of all the mathematical objects possessing that property. If this idea had been valid, every mathematical definition could have been reduced to a proposition asserting that a property is true, the statement of that property involving only the basic constituents of the language of logic. There would have been no need for specific mathematical axioms, since any proposition could have been chosen as a possible axiom (after checking that the set it defined was not empty). There would be no specific content of mathematics and the theorems would have only repeated the definitions under another guise. Wittgenstein shared that conception when he reduced the content of mathematics to tautologies.

The story of Frege's tragic failure has been narrated many times, and particularly well by Van Heijenoort (1967). It occurred when Bertrand Russell noticed that if one considers as a property of a set the fact that it is not a member of itself, the existence of the set of all the sets possessing that property is self-contradictory. This paradox and a few others meant that specific axioms were needed to avoid their occurrence in set theory. The question whether these axioms still belonged to pure logic or whether they turned mathematics into an autonomous field lasted still for some time. Russell and Whitehead proposed a version of such axioms, but in spite of Russell's protestations they were not accepted as belonging entirely to logic. It seems therefore that one may consider logicism as a doctrine of the past.

## THE FECUNDITY CRITERION

The course of mathematics did not follow the stream of logicism, which soon dried up into small trickles. It was thought better to find a collection of axioms for set theory, wide enough to encompass every valuable field of mathematics and restrictive enough to keep away from paradoxes. This task was accomplished by the Zermelo-Fraenkel axioms and we discussed some of their aspects already in chapter 3. We encountered particularly the famous "axiom of choice," which was introduced by Zermelo and raised many misgivings. As already mentioned, it refers to a set $E$ whose elements are sets. Then one defines another set $U(E)$, the union, whose elements are all the elements of the elements of $E$ (the existence of $U(E)$ is a separate axiom). According to the axiom of choice, all the sets that are elements of $E$ are supposed to be nonempty and disjoint. The axiom then states that $U(E)$ includes at least one subset $M$ that has one and only one element in common with each element of $E$. Its name means, intuitively, that one can always pick up simultaneously, or "choose," a definite element in every set in $E$.

It is very helpful to consider Zermelo's answer to some of the objections that were raised against his proposal—particularly by Peano.[1] He said first that the axiom is "self-evident," as shown by the fact that many mathematicians used it previously at one time or another, in one form or another. But because intuition is a fallible guide (remember Frege!), its true

[1] See Putnam's discussion in the book by Tymoczko (1998), pp. 54–55.

justification lies elsewhere: the axiom is *necessary for science*, because many useful results would break down if this basis were to fail.

One may check today the value of this final argument by opening a modern treatise of analysis (for instance, Dieudonné 1972) and looking at the loss it would suffer if one axiom or another, and particularly the axiom of choice, were removed. The hecatomb would be terrible! Hilbert spaces and distributions would be among the most prominent victims, and I quote only two fields with relevance to physics. Intuitionists (of whom more will be said) might say that it does not matter which parts of mathematics collapse, if what remains has the beauty of unassailable truth. They would also argue that the importance or the interest of a mathematical theory is a matter of taste, an aesthetic option, because most of them, except Hermann Weyl, did not care much about physics.

I propose to conclude, on the contrary, that the criteria deciding what is legitimate or not in mathematics are not intuition, aesthetics, or a direct relation with logic; *the essential qualities of mathematics are consistency and fecundity*. They are also the two main features necessitating an explanation.

## Formalism

During the second half of the nineteenth century, mathematicians felt again a longing for the secure Greek construction of mathematics, and they were urged to return to foundations by the dazzling variety and abstraction of new concepts. This evolution is described in many books, among which Bourbaki's *Eléments d'histoire des mathématiques* (1960) is particularly instructive.

Several fields of mathematics seemed vague and sometimes two faced concerning the exact subject of their work. When was one investigating, for instance, an $n$-dimensional linear space, and when was one considering a system of $n$ linear algebraic equations? The idea that the mathematical objects do not require a specific "interpretation" began slowly to emerge. In his inaugural lesson (1857), Riemann did not speak of "points," but of "specifications" belonging to "$n$-times extended manifolds," in which the metrical relations "can only be studied for abstract quantities and represented by formulas; under some conditions, some of them may have a geometrical representation and it is then possible to express the results of a calculation in a geometric form." In other words, mathematics is concerned by relations and not by objects. Its foundation must reflect this noncommitment.

The enterprise of axiomatization began really, first in algebra and then in classical analysis, with a foundation of real numbers and complex numbers on the set of integers. Dedekind then defined a complete system of axioms for arithmetic (which is better known when attributed to Peano, who stated in addition the principle of recurrence). The most difficult project was of course the axiomatization of set theory.

The framework of this vast enterprise was called "formalism." It relies on a conception of mathematics relying on the basic rules of logic and specifying a field through a definite set of axioms. This vision, which was hailed as unequaled by Bourbaki, gained simultaneously the rank of a dominant philosophy of mathematics and of a standard rule of the mathematical game. Many important questions came to the forefront on this occasion.

There was the question of consistency of a set of axioms, which was raised in 1900 by Hilbert in the case of arithmetic (it was the second problem in his famous lecture at the international congress of mathematics of that year). "Consistent" is in that case a synonym for non-self-contradictory. When he stated the problem, Hilbert introduced a new momentous principle: Whereas in traditional logic, the noncontradiction of a concept only made it "possible," he defined self-consistency as equivalent to the "existence" of the concept in the axiomatic framework.

The development of formalism was dominated by Hilbert's personality until about 1930, and it became rather

183

involved, at least too much so to be properly described in the nontechnical spirit of the present chapter. Only a few headlines will therefore be mentioned:

1. There was the theory of proof. Does this theory belong to mathematics, in which case a proof should be entirely formal? Or does it belong to metamathematics, i.e., to a superstructure outside mathematics? An ascetic rigorous formalism can be avoided in the second case and mathematical constructions as well as their communication become easier (one can use, for instance, a language closer to the ordinary one in a proof).

2. There was the question of independence within a system of axioms. It asks whether an axiom is not perchance a theorem resulting from the other axioms. The axiomatization of topology was particularly delicate from this standpoint.

3. There was the essential problem of decision (*Entscheidungsproblem*), to which we shall return soon.

4. The question of mathematical structures came out later under the impulsion of a group of young men choosing to call themselves collectively Bourbaki. Their aim was essentially a more economical axiomatic architecture of the edifice of mathematics. It asks, for instance, which field relies on the axioms of a previous one and needs a minimal number of further axioms to be defined. The structures of general set theory, algebra, basic analysis, and topology were then used as the main pillars of mathematics. The resulting pattern is a hierarchical construction with command lines pointing from the more fundamental to the more specialized structures or substructures. This emanation of the formalist school bears the name of "structuralism" and it has been influential in wider circles of philosophy.

One must distinguish two different aspects in Hilbert's enterprise when assessing its importance and meaning. There is first the question of method. Hilbert regarded formalism as the most convenient (and in his opinion the only valuable) method for building mathematics on a clean (and in his opinion unique) foundation. This approach was later corroborated in two different directions. Axiomatism was very successful in organizing coherently almost all of mathematics, when applied by Bourbaki with the help of structures. That was only axiomatism, however, and not strict formalism, since Bourbaki's *Elements of Mathematics* was written in a nonformalized language. True formalism found its most fruitful domain later with the development of computer science.

The other aspect of Hilbert's endeavor can be described as a dream, or one of the most ambitious philosophical enterprises of the human mind. His final program was to establish explicitly the logical consistency of mathematics, and to begin with at least the consistency of arithmetic. If the project had succeeded, it would have meant that mathematics *exists* by itself without any reference to physical reality. The undeniable existence of this entity would have practically validated, by the same token, the Platonist thesis of a reality specific to mathematics and external to this world.

It is well known that Gödel's incompleteness theorem sealed this issue by establishing the existence of nondecidable mathematical statements, which can never be proved true or false by means of a formal proof. This result fortunately had no consequence for the validity of formalism as a method. As the later proponents of quasiempiricism said, Gödel's theorem put an end to a dream, but not to anything credible. The dream was to *establish* Platonism, but its downfall did not mean that the thesis of Platonism became untenable as a metaphysical position. Gödel himself maintained it till the end of his life.

## INTUITIONISM

These developments shocked a number of mathematicians and we already evoked their perplexity with the axiom of choice. The reaction was called "intuitionism." It was led by

Jan Brouwer and represented a sort of commonsense response to the supposed excess of formalism. This point of view, which was at least partially shared by Henri Poincaré and Hermann Weyl, also had several variants into which I will not enter. It seems fair to recognize a common denominator among its proponents, which was an agreement on the ultimate foundation of mathematics on the existence of integers. There was also a less universal allegation of the necessity of an explicit construction for asserting the validity of a mathematical object, with a refusal of its existence on the sole ground of noncontradiction.

## COMPUTERS AND FOUNDATIONS

Many concepts and methods belonging to the philosophy of mathematics became almost matters of practice with the advent of computers, which behave like logical machines and can provide in principle a model for any sequence of mathematical symbols. Formal axiomatic systems, which had been introduced by Hilbert as a keystone in the philosophy of mathematics, could then be defined in terms of computer programs. The opposition between formalism and intuitionism lost some of its meaning, since a computer is also in principle a machine performing a series of arithmetical operations on integers.

This reformulation of the foundations of mathematics, or at least a part of them, was mainly the work of Alan Turing. The only price he had to pay for turning a machine into a philosopher of mathematics was to replace actual computers, as we now know them, by an ideal one without any material constraint. This "Turing machine" is supposed capable of reading an arbitrarily long input (including data and program) and printing an arbitrarily long output.

When reporting some results that were obtained with this approach (Chaitin 1982), I will look for simplicity rather than rigor, so that I will prefer simple notions to elaborate ones that would be more adequate from a punctilious standpoint. For instance, I will pay no attention to the differences occurring in the length of an output or in the duration of a calculation when one uses different types of sequential computers or different Turing machines. Rather than speaking of a sequence of mathematical symbols, I will also consider that, after writing a program in terms of the usual bits 0 and 1, the input (data and program) can be interpreted as a number.

Turing used his ideal machine to settle Hilbert's decision problem, which we already mentioned. One of Hilbert's key requirements for a formal system was the existence of an objective criterion for deciding if a proof (i.e., a sequence of symbols using the language of the system) is valid or not. When recast in the framework of the Turing machine, this question became the so-called halting problem, which asks: Is there an effective procedure (an algorithm) by means of which one can test a program to decide in advance whether it will ever halt when run on a Turing machine? Turing's answer was negative: there is no such algorithm and Hilbert's decision problem is therefore unsolvable.

Among the concepts originating from the Turing approach, algorithmic complexity is particularly worth mentioning. This notion has to do with a very large number $X$, whose complexity $C(X)$ is defined as the length—in bits—of the shortest program yielding $X$ as an output. This idea is quite different from the relation of a number to the figures expressing it in the decimal or the binary system. There are, for instance, very short programs for computing a fixed (though very large) number $N$ of decimals of $\pi$. Every sequence of these decimals therefore has a small complexity and one may say, grossly, that the complexity of $\pi$ itself is small. Conversely, a large number $X$ whose $N$ bits result from a quantum random process has a complexity $C(X)$ practically equal to $N$, since the shortest program generating it consists in using $X$ itself as a datum, and reducing the program to the order "Print."

Gödel's incompleteness theorem becomes much clearer in this approach, as shown by Chaitin. It follows directly from another theorem in information theory, stating that no algorithm can compute the complexity of every given number X. Since we are concerned with the importance of Gödel's theorem in assessing new attempts at the philosophy of mathematics, we shall listen to Chaitin's opinion on this topic:

> At the time of its discovery, Kurt Gödel's incompleteness theorem was a great shock and caused much uncertainty and depression among mathematicians sensitive to fundamental issues, since it seemed to pull the rug out from under mathematical certainty, objectivity, and rigor. Also its proof was considered to be extremely difficult and recondite. With the passage of time the situation has been reversed. A great many different proofs of Gödel's theorem are now known, and the result is now considered easy to prove and almost obvious. . . . And it had no lasting impact on the daily lives of mathematicians or on their working habits; no one loses sleep on them any more.

## Platonism

It may be convenient to insert here the question of ontology, with those of Platonism. Mathematical Platonism is a metaphysical assertion according to which there really exists a self-contained entity outside of space-time, unphysical, immaterial, which is explored and charted by human mathematics just as natural sciences investigate and chart the physical reality. It will be convenient to call this entity *Logos* in the present section. The opposite of Platonism is sometimes called constructivism, which considers mainly that the construction of mathematics is a purely human invention, while its most dedicated proponents refuse anything displaying some suggestion of metaphysics. Mathematics, if they were to be believed, is nothing but a game of free assumptions, where the fun of deduction measures the interest of the show.

Platonism, which is also called mathematical realism, always had many proponents among mathematicians and one may quote particularly in recent times (and in alphabetical order) Alain Connes, Hao Wang, Kurt Gödel, Roger Penrose, Hilary Putnam, Abner Shimony, and René Thom. Reuben Hersh said humorously that the typical working mathematician is a Platonist on weekdays, when he or she is doing mathematics, and a formalist on Sundays when writing a paper.

Most Platonists assume the existence of a close relation between *Logos* and the physical world, which goes in a one-way direction: *Logos* is supposed to govern physical reality, whereas reality has no influence on it. Putnam's position is original when he identifies *Logos* explicitly with all the possible appearances of physical reality, which he distinguishes from actuality (Putnam 1975). He justified this assertion by modal logic, a version of logic rejecting the principle of the excluded middle and allowing three truth values for a proposition—namely, "true," "false," and "possible"—with a maximal extension of the notion of possibility in the present case.

One may notice that there are different versions of *Logos* among Platonists. Gödel, for instance, considered that the concepts of set theory are the building stones of *Logos*, whereas Thom was nearer to Plato when he included almost everything in mathematics as representing some aspect of *Logos*. Putnam made it the home of possibilities and Connes sees it in close relation to the physical laws. This variety of views on *Logos* may even be extended a little further to include "ultrafinitism"; this ontological position originated recently among the community of computer scientists and it assimilates the universe to a gigantic computer, which would actualize the possibilities, laid open to reality by the laws, in every round of its calculations.

The arguments in favor of Platonism rely essentially on the fecundity and consistency of mathematics. Fecundity is counterbalanced by the resistance of mathematics to the will of its creator or discoverer, and the proponents of Platonism often invoke that feeling. It may be noticed that the best mathematicians are particularly tempted by a belief in Platonism or an inclination toward it, probably because they appreciate better than everybody else the vastness and the consistency of their science, and they experience directly how it seems to guide them through its refusals.

There is, however, an even stronger argument against this point of view, which asks how the mind of a human being could establish a contact with a reality lying outside space and time. The only answers are more or less close to religion and, while they greatly satisfy some people, they also make some others more careful about the implications of Platonism.

## THE COGNITION APPROACH

Some biologists have called attention to the brain abilities that are used in mathematics, and Jean-Pierre Changeux, for instance, has replaced this question within the framework of cognition sciences (Changeux and Connes 1995). These abilities are then considered as a product of evolution, over many millennia, and they show no essential difference from many other human mental activities. Abstraction, the appeal of problem solving and intellectual performance, or the attraction of general vistas, are frequently met in mankind and their application to a specific field does not require an explanation. This approach is particularly interesting for the link it establishes with the origin of reason and, as a matter of fact, we used it in this book when we discussed abstraction and classical reality in chapter 2. We shall see its main limitation later.

## QUASIEMPIRICISM

A new approach to the philosophy of mathematics was introduced by a number of mathematicians and philosophers in the 1970s and was called "quasiempiricism" by Lakatos and

Putnam. An anthology of relevant texts, collected by Thomas Tymoczko (1998), provides so good an introduction to this trend of thought that I could find no better account of it than sewing together various quotations from Hersh, Lakatos, Putnam, Goodman, and Tymoczko himself, and adding a few comments.

## The Source of the Philosophy of Mathematics is in Their Practice

According to Hersh, the great period of foundational controversies, which extended from Frege and Russell to Brouwer, Hilbert, and Gödel, had as its aftermath an impasse in mathematical philosophy. It is one thing to accept the assumption that mathematics is a source of indubitable truth when, like these people, one believes that the foundation is nearly attained. But it is quite another thing to go on accepting it to shape one's philosophy, long after any hope of attaining that goal has been abandoned. What one needs is a new beginning and it must start from the ongoing practice of mathematicians. Over the last two-thirds of a century, foundational research and ordinary mathematical practice evolved along different lines and, in order to revive the philosophy of mathematics, one must return to its source for a fresh look.

## On Two Kinds of Theories

According to Lakatos, there are two kinds of mathematical theories: the Euclidean and the quasiempirical ones. In a Euclidean theory, the basic statements are contained in axioms and the rules of inference are precisely determined. Truth is injected into the system at the level of the axioms and it "flows downward" to their deductive consequences. Knowledge, when it results from a proof, is considered infallible. Quasiempirical theories begin, on the contrary, while their subjects are still indeterminate. They can describe and manipulate many variations of an idea and their goal is to get at the underlying principles.

Knowledge is admittedly fallible. The basic statements consist in a set of assumptions, observations, or conjectures, such as may come from computations. The rules of inference are sometimes not very precise. Truth and falsity are injected into the basic statements, but in quasiempirical theories truth does not flow downward; falsity, on the contrary, flows upward. The axioms or basic principles of quasiempirical theories are thus usually the outcome of some bold speculation having survived a severe criticism. Most mathematical theories are considered by Lakatos as belonging to the quasiempirical type or having passed through it; Putnam points out that established principles such as the correspondence between the real numbers and the points on a line or the axiom of choice have been the outcome of quasiempirical methods.

As a comment, one may add that the description of quasiempirical theories is strongly reminiscent of theoretical physics. It is much easier, as a point of fact, to point out quasiempirical parts of mathematics by looking at the mathematics of physics than at the so-called "pure" mathematics (except if one goes back to their past history). One may quote then as examples the sums on Feynman histories (which have now, by the way, invaded vast fields of pure mathematics) or the manipulation of infinities in renormalization theory.

### Mathematical Truth Is Fallible

This is another leitmotiv of quasiempiricists. Many of them say that, if one looks at mathematical practice with an unprejudiced eye, one observes prominent features that were ignored by traditional philosophy. They assert particularly that mathematical knowledge is inherently fallible and that no foundation can make it infallible. Many arguments are brought in support of this thesis, because it obviously breaks with the traditional prejudice in favor of an unquestionable truth of mathematical reasoning. Lakatos gave long lists of errors, which appeared in more or less famous works and were only corrected long after. He almost made fun of the contention that truth might be ensured by a formal proof. He also stressed that, on the contrary, almost no proof of an important theorem is written in a formal language and, even if it were, it could be considered as a computer listing and one would have to worry about bugs. The number of steps between the axioms of set theory and an important theorem in functional analysis is quoted as an example of so long a sequence that the belief that they have been actually checked is illusory.

Hersh had a significant remark on that topic, which sheds much light on the quasiempirical conception of truth while opposing it to formalism, when he said: "It is understanding that verifies the correctness of formal computation, not only the other way round." René Thom made similar statements and he also called attention to the importance of analogy in mathematical thinking.

## Mathematics versus the Natural Sciences

Many proponents of quasiempiricism insist on a close relation between mathematics and the natural sciences. This is probably one of the reasons why they insist so much on the fallibility of mathematics, which sometimes looks more like a provocation than anything worth believing (if they think so, why are they doing mathematics, except for the self-seeking pleasure of a game?). There is no doubt that the activity of research follows the same patterns in mathematics and in the abstract parts of the natural sciences, some people saying that the difference shows up only at the time of writing a paper.

The deep analogy between mathematics and theoretical physics will be discussed in the Appendix to this chapter, particularly where consistency is concerned, but one cannot avoid the fact that many crucial statements in physics find their validity in a falsification test. Nothing analogous exists in mathematics, which is why it is not an empirical science but at best a quasiempirical one. Lakatos made several proposals for finding "falsifiers" of mathematics, but they were very poor candidates for the role of guardian of truth. Finally, quasiempiricism falls back on the question from which the foundation crisis started: What are the source and the criterion of truth in mathematics?

One may notice that quasiempiricism insists on mathematical practice, which is essentially a public activity. This is a point where it is at odds with the standard foundational attitude, holding that mathematics is essentially a private affair, which takes place in a person's mind, public practice being only an external symptom of it. The new emphasis on mathematical practice brings with it an emphasis on "the mathematical community as the ultimate source of mathematical activity."

This last statement is borrowed from Hersh, who is not far from replacing the word "activity" by "truth," though he does not. He is then on the verge of positivism, for which the ultimate criterion of truth is a gentleman's agreement between knowledgeable people. That brings up the familiar controversy between realism and positivism, which is not new in science and has been particularly acute in physics since the beginnings of relativity and quantum mechanics. Positivism, on the other hand, is a newcomer in mathematics, where it appears in new-fashioned garments. Nicholas Goodman, for instance, believed he could embrace positivism and realism simultaneously through a curious twist to the notion of reality (Goodman 1979). He stated the following axiom: "Anything that is practically real is objectively real." Roughly speaking, this principle of "objectivity" boils down to the following statement: "If a concept $X$ plays an important role in a theory, and if failure to acknowledge the role of $X$ severely limits the theory, then $X$ is practically real. Moreover, in the absence of a strong argument to the contrary, the presumption must be that anything practically real is objectively real."

Some proponents of quasiempiricism go back to Platonism in one form or another, as we already mentioned in the cases of Thom and Putnam. But in either Hersh's vague proposals,

Lakatos' search for falsifiers, or Goodman's rewriting of positivism, one can perceive a feeling of dissatisfaction. Goodman best expressed this impression when he described the philosophy he dreamed of: "Such a philosophy of mathematics would be only one chapter in a larger philosophy of science. That philosophy would make it clear in what sense there is only one objective world and how it is that the objects studied by the mathematician, many of which are not related with physical reality, can nevertheless be seen as parts of that world. Unfortunately, that philosophy has yet to be formulated." I am tempted to read that longing for a better philosophy as meaning that physism should be now at last introduced and discussed. That is done in the next chapters.

## APPENDIX. THE METHOD OF PHYSICS: A COMPARISON WITH MATHEMATICS

The analogy between the methods of physics and mathematics—at least from the standpoint of quasiempiricism—is probably most apparent in the four-stage method, which was proposed in a previous book as the most convenient description of the modern construction of physics (Omnès 1999a). This method brings together several aspects of the scientific method, as stressed by Einstein, Feynman, and Popper. I proposed to call it the "four-stage method." This name does not suppose that the process of research always follows the same pattern, going invariably through four stages that succeed each other in the same order. It is meant only to stress that four distinct acts of experiment and thought enter at one time or another in the construction of a theory and the recognition of its truth. One might speak of components of the method rather than "stages" or "steps," but that does not matter much.

Through the first step of the method, the *empirical* one, a science establishes the first contact with its object. This preliminary work consists usually in collecting many facts by

experiment and observation. Some facts are sometimes summarized in empirical rules, expressing that some regularities
have been discovered although one does not know yet, usually, why the rules hold. A famous example in astronomy was
the observation of planetary motions by Tycho Brahe, culmi-
nating in the discovery by Kepler of three important empirical
rules. The history of quantum mechanics is full of such discoveries and temporary rules, such as, for instance, Bohr's
model of the atom, the first version of Pauli's principle, the
spin of the electron, isotopic spin, strangeness conservation
and SU(3) flavor symmetry in strong interactions, parity violation, the existence of the second and third families of quarks
and leptons. The list could be very long.

The second step of the method is fascinating, particularly
when it brings a science to a high formal level. It may be called
the *conceptual* stage. It is essentially an act of pure conception,
invention, or imagination as one prefers to call it. Its aim is to
conceive the *founding concepts* of a theory, which will be able to
organize the facts and to guess the form of the laws. This step
occurred in the case of classical mechanics when everything
was yet to be invented, including the scientific method itself.
Great pioneers such as Galileo, Descartes, Fermat, Pascal, and
Huygens worked at it until the enterprise was achieved by
Newton. The basic concepts were then those of mass and force,
absolute space and time (from which one could infer position,
velocity, and acceleration). There were only three principles: the
principle of inertia (the uniform straight-line motion of a body
experiencing no force), the equality of action and reaction, and
the "fundamental principle of dynamics" relating force, mass,
and acceleration.

It took time to realize that the basic physical concepts and
the formulation of the laws of a theory do not result from a
standard induction process, but from an invention, a creative
act of the mind pondering on problems arising from the facts.
The search for consistency is central in this endeavor. One can
see the conceptual stage at work in many places during the

nineteenth century, but its character became manifest only with relativity and quantum mechanics. It will not be necessary, however, to review many examples, and what we encountered already with Heisenberg matrices, the Schrödinger and Dirac equations, or Feynman histories should be sufficient. We might write a long list of occurrences of the conceptual process, which became more and more clearly mathematical as quantum theory extended its range and reached deeper layers of the world of particles.

The third step of the method squeezes the juice from the principles. It is an intellectual exercise looking for the consequences of the supposed principles. One may call it the *elaboration* stage (elaborating the content of the principles). There are many examples in classical mechanics, such as the discovery of Neptune, Foucault's pendulum, the laws of elasticity and of fluid mechanics deriving from the Newtonian principles. Nowhere is the work of physicists closer to the practice of mathematics: one conjectures a consequence of the principles, as one does in math for a theorem resulting from a given system of axioms. One proves this consequence more or less as in mathematics, although one must check its consistency further with already known data, and precise predictions generally involve much computing. The end of the process is always something like saying: such and such properties are consequences of the supposed principles and they are true if the principles are right. Jean Petitot has described the principle of elaboration as the construction of a virtual reality, which is finally compared with empirical reality, in the fourth step of the method.

This last step is concerned with *falsification*. This is the crucial moment when a theory, still only a hypothesis, is offered for rejection and runs the risk of being shown false. The predictions that were drawn from the principles through elaboration are submitted to the test of experiment. There are two possibilities at this critical moment. The expected consequences are found correct (all of them) and then the theory becomes an acknowledged science, or some consequence is disproved and the theory is deemed worthless. This is how the phlogiston and cold fusion were falsified and how general relativity was verified when curved light rays were observed.

Finally, the scientific method is entirely controlled by experiments; at the beginning when empirical data are collected and at the end with falsification. The conceptual stage is also found in mathematics, where it is perhaps even more mysterious, like creativity itself. Poincaré was fascinated by its psychological aspects, which he described at length in his books. The conceptual act shows no significant difference when it is concerned with a purely mathematical concept or a new setting for a natural law. There would be much to say about its relation to pattern recognition (which means, by the way, that pattern recognition could have a much wider scope than usually envisioned in the framework of artificial intelligence). The conceptual stage seems in any case free from any a priori constraint, probably a necessary condition for leading human knowledge as far as deeper laws based only on pure freedom.

Finally, the conceptual and elaborative stages show absolutely no difference from the corresponding steps in mathematics, except of course for the specific subjects of research. The central role of consistency is the same, and the fact that the physics and mathematics communities do not pursue it exactly in the same way is only a question of topic and culture. The fact that physics can rely on experiments sometimes makes the theoreticians a bit more careless than if they were doing pure mathematics.

As a conclusion, we may say that there are *two falsifiers* in the natural sciences: consistency and experiment. Mathematics has only one, consistency, and this is why it is so demanding in this respect.

# Physism: The Thesis

**W**e are now reaching the end of our journey, with the statement and discussion of physism. According to this proposal, mathematics belongs essentially to the laws of nature, and most closely to those of physics. The present chapter asserts the thesis and it proceeds as follows. The historical development of ideas is first recalled. Then one compares the laws of nature with mathematics, to show how small their difference is. Their greatness, or the two "miracles" of their existence, is then put into a common perspective. Some lesser clues for physism are next added, before proceeding to a unified discussion of the problem of consistency. The discussion, pro and contra, of the thesis in a philosophical perspective and its relation to other approaches to the philosophy of mathematics will follow in the next chapters.

## HISTORICAL APPROACH

Much of what will be said in this section was already mentioned in previous chapters, but it is probably worth recalling. Math and physics sailed for a long time in company or under the same command: the Greek *Logos*, God's will, or some universal harmony. Then came three crises, one in mathematics and two in physics—because of relativity and quantum mechanics—with a confusing discordance in timing.

Mathematics ran away from nature—or at least classical reality—when apparently unnatural notions, such as crazy nondifferentiable functions, Lebesgue integral, infinite sets of various sorts, and highly abstract structures were found necessary for its consistency. Most mathematicians and philosophers took it then as a thing in itself, or at least an isolated continent of knowledge, which might happen to find applications in other domains.

When looking for the early roots of physism, one must mention Poincaré and Hilbert as its two forefathers. Poincaré valued intuition highly and he cleanly distinguished the essence from the technique in mathematics, the essence being close to the universal laws of nature while the techniques, though necessary for proof, are not on the same footing. He perceived that physics was undergoing a great change in his time, to which he contributed: he was a cofounder with Einstein of special relativity and he made a significant contribution to the early quantum theory.[1] Hilbert's contribution to general relativity is also important,[2] but much more important is his vision of theoretical physics as an axiomatic and demonstrative mathematical construction, which is undoubtedly at the core of physism.

We mentioned in chapter 3 (the section "A Parallel with Physics") how the problems of interpretation in quantum mechanics shed doubt on the objectivity of the theory. That was certainly important in widening the apparent chasm between the foundations of physics and of mathematics among philosophical circles during the twentieth century. Had it been proposed some decades ago, physism would have appeared naive and

[1] Poincaré proved that Planck's law for the black-body spectrum necessarily implies discrete energies.

[2] Hilbert gave a formulation of Einstein's laws relying on the least action principle, and this approach is much used nowadays.

untenable.[3] Perhaps I am too naive and I overvalue the present return to objectivity, but I think that the perspective has changed drastically and one should look with a new eye at many clues that appeared throughout the last century, which will now be considered.

## A HOST OF COINCIDENCES

Still from the standpoint of history, the progress in mathematics and theoretical physics was strongly related for a long time, one science suggesting problems and answers to the other, in both directions. A few interactions of that kind, although well known, are particularly worth mentioning for their significance in the present context.

- Partial differential equations appeared first in physics, with vibrating strings, sound waves, and heat transfer. Fourier series were then invented for solving them. The study of these series (and some of their modifications) led to the construction of nondifferentiable functions and started Cantor's investigation of infinite sets.
- The origin of vectors is an intricate combination of physics and mathematics. Various integral formulas (by Green, Ostrogradsky, Riemann, and Stokes) were discovered in the framework of electricity and magnetism. Their generalization led later to differential forms and cohomology, but they had meanwhile been a key in the discovery of Maxwell's equations in electrodynamics.
- Tensors were first introduced in elasticity. When formulated

---

[3] An interesting milestone is the article by Wigner (1970), "On the unreasonable efficiency of mathematics." He was a great physicist and a careful mathematician, but he was also strongly impressed by the difficulties of interpretation since he was one of the main proponents of the collapse of the wave function through the consciousness of an observer.

mathematically, they contributed to the development of differential geometry and the general theory of curvature, and led to the representation of Lie groups.

- Lie groups were initially a mathematical notion, suggested by geometry as a generalization of finite groups. When Poincaré applied them to the invariance group of Maxwell's equations, he obtained most key results of special relativity, independently from Einstein.
- Felix Klein had shown previously the close relation of different geometries (Euclidean, projective, and so on) to their symmetry groups. Algebra had led on the other hand to a powerful theory of quadratic forms. It was then easy for Minkowski to recognize the basic mathematical structure of Einstein's theory of special relativity as a four-dimensional space-time, whose symmetry group leaves invariant a non-positive quadratic form. This geometrical interpretation yielded, a few years later, the basic idea of general relativity.
- Geometers felt themselves in a no-physics land when they invented non-Euclidean geometries, $n$-dimensional spaces, and the connections on Riemannian spaces. But that was exactly what general relativity needed.

Quantum mechanics and modern mathematics displayed still richer and long-range interactions, of which I mention a few.

- Matrices were ready for Heisenberg's formulation of quantum mechanics, just as the Sturm-Liouville spectral theory of differential operators was an ideal tool when Schrödinger elaborated the consequences of his equation.
- Hilbert spaces, on the contrary, were almost mathematical curiosities with few applications when Von Neumann developed the spectral theory of the linear operators in these spaces, for the purpose of quantum mechanics. The completeness of a Hilbert space of functions, essential for many properties of quantum wave functions, relied essentially on Lebesgue's

integral. Not so long before, this kind of integral was considered by physicists as a typical example of the playthings with which mathematicians were amusing themselves uselessly. On the other hand, Dirac's kets and bras, which had initially looked senseless to the mathematicians, were identified by Frederic Riesz as vectors and linear forms, and that was one of the main starting points of duality theory in analysis. From there on, Hilbert spaces and their generalizations (particularly Sobolev spaces) became central concepts in pure mathematics.

- Dirac had invented another tool, the delta function, to get a simple description of continuous spectra, but it remained suspicious for a long time to supercilious mathematicians. Twenty years later, however, it entered through the main gate in Schwarz's distribution theory, which relied, by the way, on the application of duality to the case of locally convex spaces (Fréchet spaces).
- Not so long ago, partial differential equations were still considered by some influent mathematicians as some foreign pebbles in the main stream of mathematics.[4] Linear partial differential equations returned to the sheepfold only recently with microanalysis, a beautiful construction involving differential geometry and distribution theory. It is slowly penetrating the physics community and already provides the most convenient methods for deriving classical mechanics from quantum mechanics.
- Elie Cartan's study of semisimple compact Lie groups provided a mathematical background for the theory of spin. Their classification, again by Cartan, was also found essential when physicists were trying to get some order in a plethora

---

[4] I may mention a conversation with Alexandre Grothendiek in the early 1960s, during which he asserted that partial differential equations "do not belong to mathematics." He was probably impressed by the then recent discovery (in 1956) of linear equations, where all the data are infinitely differentiable functions and they nevertheless have no solution.

of particles in the late 1950s. Gell-Mann's discovery of what is now called flavor SU(3) symmetry was an essential step in this enterprise, which led him to the beautiful and deep idea of quarks.

- Wiener's continuous integral and Feynman's sum over histories are closely related, the second one embodying the first (when statistical mechanics or imaginary times are introduced). Wiener's construction belongs, however, to the main (axiomatic) stream of mathematics while Feynman's does not. It stands rather as the most beautiful example of quasi-empirical mathematics I know, still flouting formalism since it is essential to the standard model of particle physics and has already yielded magnificent results in pure mathematics.

- Hurwitz invented fiber spaces as a mathematical structure, now central in differential geometry. Physicists and mathematicians together were, however, extremely surprised by the discovery that fiber spaces stand at the mathematical foundations of the gauge theories in the standard model. There are fascinating consequences: For instance, the connection forms on the relevant fiber space become the potential vectors describing the spin-1 intermediate bosons of physics (the photons, the $W$ and $Z^0$ bosons of weak interaction, and the gluons storing interactions). Measurable fields (for example, the electric and magnetic fields) are associated similarly with the curvature of the fiber space. The arbitrariness of connection forms on the fiber space corresponds to the arbitrariness in the choice of a physical gauge (for example, the freedom in the choice of the potential vector in electrodynamics). The algebraic expression for the covariant derivatives on the fiber space completely determines the interaction (coupling) between the particles.

There are many other examples in experimentally confirmed physics, and specialists think a lot of many others in string theory and noncommutative geometry, but I am not well aware of the latter. It should be clear anyway that physics and mathematics have been closer than ever during the twentieth century, even though mathematicians and physicists worked independently most of the time. What would a jury say of the culprit and the defendant when so many coincidences link them up? No doubt: it's the same person! What would a geographical society conclude when the maps, brought back by two explorers, show so many identical features? No doubt: they explored the same country! But there are still more clues, so let us look at them.

## Is There a Significant Difference between Mathematics and Theoretical Physics?

"Can one see a difference when a mathematician is thinking or when he or she is sleeping?" asked André Lichnerowicz. It is surely not much bigger than the difference between mathematics and theoretical physics, which is not much bigger than the hair on an egg, as the saying goes.

There is one obvious difference, of course: The people belong to two distinct communities and they are not lodged in the same university building. It may be noticed, however, that the increasing sophistication of physics has split its community into experimenters and theoreticians during the last century. One can no longer say of anybody, as was said of Fermi, that experimenters believed him to be one of them, and theoreticians too. The new group of theoreticians can be distinguished from the bulk of mathematicians by some difference in culture, but surely no greater than the differences inside each body itself. Some of their topics of interest are not given the same importance; very few mathematicians work, for instance, on the analysis of experimental data, and no physicist publishes papers on fundamental logic. But let us not pursue this splitting of egg hairs.

The appendix in the previous chapter indicated a deeper origin for the specialization of the physics community into experimentalists and theoreticians: It is a matter of method,

205

which is partly dictated by the importance of empirical data and experimental falsification, whereas theory is quite close to mathematics. We noticed particularly that the conceptual and the elaboration stages in the method of physics coincide with two similar aspects of the work in mathematics, and this comparison warrants a few more comments.

Both mathematicians and physicists put much weight upon consistency, but this is so important a topic that we shall deal with it by itself in a moment. There is also the question of beauty. Great mathematical constructions and smaller sparkling jewels are often praised for their elegance and beauty, and much of the attraction exerted by mathematics on its devotees is due to that. But in an insightful paper, Dirac (1963) wrote, on the other hand:

> It is more important to have beauty in one's equations than to have them fit experiment. . . . It seems that if one is working from the point of view of getting beauty in one's equations, and if one has really a sound insight, one is on a sure line of progress. If there is no complete agreement between the results of one's work and experiment, one should not allow oneself to be discouraged, because the discrepancy may well be due to minor features that are not properly taken into account and that will get cleared up with further developments of the theory.

Dirac also gave in the same paper an account of the relation between physics and mathematics that is almost physism:

> It seems to me one of the fundamental features of nature that fundamental physical laws are described in terms of a mathematical theory of great beauty and power, needing quite a high standard of mathematic for one to understand it. You may wonder: Why is nature constructed along these lines? One can only answer that our present knowledge seems to show that nature is so constructed. We simply have to accept it. One could perhaps describe the situation by saying that God is a mathematician of very high order, and He used very high mathematics in

constructing the universe. Our feeble attempts at mathematics enable us to understand a bit of the universe, and as we proceed to develop higher and higher mathematics we can hope to understand the universe better.

## *The Tricks of the Trade*

When one turns from these higher considerations to daily practice and compares the tricks of the trade in mathematics (as met in chapter 3) with those of theoretical physics, one has difficulty in finding a difference. I shall again give a few examples.

- The quest for deeper axioms in mathematics corresponds to the search for more fundamental and more convenient principles in physics. Witness the foundations of general relativity or quantum mechanics, but see also Maxwell's laws, and the elucidation of the principles of equilibrium thermodynamics by Caratheodory (who was, by the way, a mathematician) or of the foundations of statistical mechanics by Gibbs.

- There is the quest for generalization, well known in mathematics. This topic is also so vast in physics that I shall quote only one example, in irreversible thermodynamics. Many different empirical laws accounted for different irreversible processes in the nineteenth century. They included Fourier's law for heat transfer, heat production by slow friction, dissipation of energy in a fluid through viscosity, and in the case of electric and magnetic phenomena, Joule's law for heat production by an electric current, other empirical laws for the Thomson, Peltier, Ettingshausen, Leduc-Righi effects, and there are others. Lars Onsager noticed in the mid-1930s that they could be united into a unique general formula, completely explicit—which led, by the way, to the recognition of new effects. The virtues of generality brought new insights, particularly a symmetry property among the various coefficients in the different laws, which is very useful in practice. It goes without saying that Onsager's work was generalized later in its turn.

207

- The game of conjectures, lemmas, and proofs, analyzed by Lakatos in the case of mathematics, is practically the same in theoretical physics. One could for instance write a book about its application during the history of particle theory, from Dirac's equation to the construction of the standard model.
- The trick of inverting a problem is also sometimes used in physics. It may be remembered, for instance, that Schrödinger developed his version of quantum mechanics through a process of "quantization" of classical physics, in which quantum rules were guessed from a knowledge of the classical laws. The inverse process is presently favored, since quantum mechanics has shown its universal validity, contrary to classical physics. Rigorous methods, as well as quasiempirical ones, are therefore used for deriving various aspects of classical effects from the quantum principles. There are many practical and industrial applications in chemical, mechanical, and electrical engineering, including, for instance, superconductive systems, chaotic dynamics, and nanotechnologies.

## A First Statement of Physism

How should one state the thesis of physism? One cannot underestimate a fundamental difficulty in any proposition of a philosophical nature, which is the intrinsic limitation owing to the fuzziness of the words entering in its formulation. If one considers the main theses in the philosophy of mathematics, this fuzziness is smallest in the case of intuitionism, in so far as one may pretend a good understanding of its foundation on the integers. Platonism goes to the extreme opposite direction, with a logos of which almost nothing can be said, except that its existence is supposed to calm down one's philosophical anxieties. Constructivism begins with "math is nothing but" and it escapes the difficulty, but this "nothing but" is simply the old panacea of skepticism, which is the exact opposite of the enterprise of science. Formalism was great, because it was a well-defined program within well-defined logical boundaries with no fuzziness anywhere, or in a nutshell, a marvel of philosophy. Unfortunately, it failed, and that was the prize it received for its sharpness.

Philosophers are well aware of the curse of verbal fuzziness and they try to escape it by way of an exact language, at least as exact as it can be. But the curse responds: the more precise a thesis is the sooner is the time when it is shown wrong and sent back to the fuzziness of infinite mending. Physism must also be examined in that light. When it states "$A$ is $B$," where $A$ = mathematics, the question of what $B$ stands for turns out to be tricky. We would like to equate $B$ to something belonging to nature, but one of the main outcomes of modern physics is that we cannot tell exactly what reality is. There is something obvious in it, which we called classical reality, but, surely, mathematics exceeds the limits of this part of reality. On the other hand, the great paradox of modern science is that we know the laws governing the universe and its particles better than the thing itself: reality.

Is this again a return of the curse of philosophy? I do not think so. Although we cannot, we should not, assert that we know the laws exactly, their existence has reached the status of a fact—as explained in chapter 12—so at least we may safely assume that we do. On the other hand, everything we said up to the beginning of this chapter was concerned directly with the relation between mathematics and the laws. The thesis of physism is therefore that, in "$A$ is $B$," $A$ stands for mathematics and $B$ for the laws. We then understand the two terms, namely, "mathematics" and 'the laws," but their relation "is" remains extremely fuzzy. This is therefore the point where some philosophical accuracy will become necessary.


## What For?

"Chat, chat, chat, you can only chat." This is a famous saying by a philosopher—also a parrot—in a novel (Queneau 2001).

Rather than saying "*A* is *B*," we might think of saying "mathematics is the language of the laws, and nothing but that," or "mathematics belongs to the laws," or else "mathematics is the form of the laws" (or better, as a slogan, "mathematics is the law of the laws"). As a matter of fact, I mean all these proposals as various shades or various complementary assertions of physism. A question we might consider more concretely, however, is "What for?" or for what purpose is physism worth asserting?

We may go back for that to a startling expression of Putnam, who equated the existence of mathematics with a "miracle." "Why did this miracle happen?" then became the main purpose of the philosophy of mathematics. Putnam chose a form of Platonism as his own answer, but we may leave this option aside, and concentrate on the question itself. An essential complement to Putnam's point of view is that the existence of mathematics is not the only miracle we know of. Another miracle is at least as impressive: I mean the existence of the laws of nature. Their consistency and their universality are like those of mathematics: impossible to prove, but nonetheless wonderful and practically inescapable. The main achievement of physism could therefore be both modest and priceless: *Physism does not try to explain why there are such miracles but reduces them to a unique one.* It aims in other words at a unification, as when optics and electromagnetism were unified, or electromagnetic interactions with weak interactions, just a unification of two fields of knowledge, allowing a new foundation of knowledge.

We won't try therefore to explain what mathematics *is*, except for saying that this question coincides with the unattainable explanation of what the laws *are*. As far as explanation is concerned, our only aim can then be to explain why mathematics is so fecund and consistent, given the known characters of the laws. Other features of mathematics are easier to deal with and they will be considered in the next chapters.

## Is Everything in Mathematics Concerned?

We played earlier with the idea of a super-Bourbakian construction of theoretical physics, where mathematics would be interlaced with physical concepts, the two following each other in the many sections that a consistent construction requires. The standard axioms of set theory might then appear as some sort of physical principle number one. The strong constraints of consistency would then imply that everything, or almost everything necessary, useful, or playful, in these two sciences would follow. It would obviously include arithmetic, algebra, analysis, topology, geometry, and probability calculus. The underlying axioms and concepts of basic logic would be introduced first of course (as some principle of natural science number zero), and one might designate them by the slogan of the law of the law of the laws. I will not elaborate on this game, except for stressing that *consistency* is the crux of physism. It is not explained: it stands as a prior datum of knowledge, a condition for the existence of truth.

A significant question in this approach is whether everything in mathematics is encompassed in the assumption of physism. I leave aside, of course, some useful mathematical concepts that are still in a quasiempirical state: one may think, for instance, of the sum of Feynman histories. There is no reason to suppose that they will not return to the main stream of axiomatic math at one time or another. If not, their formulation would rejoin the bulk of principle one. When looking at mathematics, the only place where the question arises whether it belongs to the area of physism seems to be higher logic. Independent systems of axioms exist, for instance, in set theory, as shown by Paul Cohen's results on the continuum hypothesis and the axiom of choice. Have these mutually exclusive mathematical theories an equal place in physism? We can't tell, no more than we can tell what future laws will be. There is at least no cogent reason for rejecting some of them into a hell of nonsense. The same is true apparently of Robinson's nonstandard analysis.[5] In any case, my own knowledge of higher logic is so poor that decency asks me to stop at that point.

## THE MEANING OF GÖDEL'S THEOREM AND THE OMEGA NUMBER

I wish, however, to return to Gödel's theorem on undecidable propositions and say a few words about Chaitin's fascinating Omega number, because they both shed an interesting light on the meaning of physism.

### Gödel's Theorem

Let us suppose for the sake of argument that Gödel's finding was that the decision problem has a positive answer. It would follow that Hilbert's program was consistent. Mathematics would then exist as an independent entity, or exist by itself, as a philosopher would say. There could be no objection to considering it as a reality.

One would then be facing again the essential difficulty of Platonism, which would then become the opposition of two kinds of realities existing by themselves, the physical one and the mathematical one. The mystery would then be thicker than the one previously encountered with nonclassical reality. Some sort of idealist philosophy, like Berkeley's, or a Leibnizian preestablished harmony, could perhaps emerge as an answer, but certainly not physism. An impressive way to express the meaning of Gödel's theorem, from our present standpoint, would then be that it acted, retrospectively, as a falsifier through which physism has gone successfully.

---

[5] An opposite opinion is expressed by Connes (2000).

## The Omega Number

Let us briefly describe the concept of the Omega number (Chaitin 1982). We sketched in the previous chapter the notion of a program on a Turing machine and the distinction between halting and nonhalting programs. Turing's theorem was also mentioned, according to which there is no algorithm for deciding a priori whether a program will halt or not. An important step in the proof of this theorem consists in showing that the set consisting of all possible programs is countable, so that it can be arranged in an infinite sequence. This set sequence is not uniquely defined because it depends on the Turing machine one is considering, but this point is unessential since one can agree on a specific machine to provide a standard.

The Omega number of this standard machine is defined as a probability in a rather roundabout way. Let us assume for definiteness that every program contains, once and only once, a sequence of binary digits equivalent to the instruction "END." One defines a random process for producing programs, which consist in drawing randomly (heads or tails) a sequence of bits 0 and 1, until one draws the equivalent of END in binary symbols. This procedure generates a correct program from time to time, which can be tried on the standard machine. Then sometimes the machine halts and sometimes it does not. The probability for the machine to halt, written in base 2, is by definition the Omega number $\Omega$. It is a perfectly well-defined number although, of course, it must be defined more carefully than I did.

Chaitin proved that $\Omega$ is not computable. It can be approximated, however, by means of a very slowly converging process. This result is not particularly surprising, but the consequences that would have followed from the—impossible— knowledge of $\Omega$ look fascinating. If this were so, every mathematical conjecture would then have received a definite answer! This remarkable statement is shown roughly as follows. Every conjecture $C$ can be associated with the program $P$ asserting it. The validity of the conjecture $C$ is then shown to be equivalent to a halting problem, which means that the program enumerating all the correct proofs of $P$ does halt. It can be shown, moreover, that a finite number of bits of $\Omega$ contain enough information to answer this halting problem and decide whether the conjecture $C$ is true or not. So, in a nutshell, $\Omega$ contains in principle every truth in mathematics—but we cannot know it.

Omega is a very strange number. Charles Bennett and Martin Gardner called it cabalistic. Nothing would forbid a human being from remembering it and using it, if he or she were told its value. It would be tantamount to knowing every secret of mathematics. But that knowledge cannot come through a rational process: it could only be the gift of a supernatural revelation!

If the Omega number had been computable, it apparently would not have been a falsifier of physism. It looks, from our standpoint, more like a falsifier of the dreams of a final theory (see Weinberg 1993). This dream assumes that there exists, in principle, a consistent set of physical principles and that this set, moreover, is unique. If so, the Omega number would contain the final principles of physics as possible conjectures. It would also hold the secret of their consistency and their uniqueness. No experiment would be needed to attain the absolute truth about this world, if one only knew $\Omega$. This is at least the assertion on $\Omega$ resulting from physism, which identifies mathematics with the form of the laws. Conversely, the noncomputability of $\Omega$ is in good accordance with physism, which considers that there are *two* falsifiers in the philosophy of knowledge, namely, consistency *and* experiment.

## THE STATEMENT OF PHYSISM

Finally, we can state the thesis of physism explicitly. We shall present it in its strongest form, which relies explicitly on axiomatism:

*There are basic axioms for logic and mathematics. These axioms are laws of physics. They are recognized through two inseparable criteria: their fecundity in the construction of mathematics and their necessity for a statement of the laws of physics. This fecundity can be explained in view of the universality, subtlety, and richness of the laws: the basic axioms must be fecund enough to allow a statement of the laws in the language of mathematics.*

*Conversely, they generate every possible field of mathematics. New laws, new axioms, new fields, are possible and they may be discovered by further research.*

*Consistency is equally necessary in mathematics and in the laws of physics, which are inseparable. Consistency cannot be explained, but it stands as one of the two criteria of truth. The other one is experimental falsification of a mathematical proposition purporting to express a law of nature.*

# Physism and the Philosophy of Mathematics

A peculiarity of physism resides in its agreement with the majority of earlier approaches to the philosophy of mathematics. Physism extracts from each of them its most significant part, the feature that made it attractive and plausible, although physism distinguishes itself at the same time from their most questionable aspects. We shall now examine the best known philosophical theories about mathematics in this light, leaving aside Platonism, however, because it is centered in ontology and this question is left for the last chapter.

## THE COGNITION APPROACH

We begin with the cognition approach, which does not acknowledge any specific ability in a mathematician, except for a conjunction of intellectual aptitudes existing in most human beings (Changeux and Connes 1995). The origin of these aptitudes is attributed to evolution, with an adaptation of the brain to the regularities of nature throughout the series of animal species leading to ours. Murray Gell-Mann and James Hartle (1991) sketched a more general approach in the same spirit with the idea of an IGUS (information gathering and utilizing system), which applies to all the forms of terrestrial and extra-terrestrial life and assumes that universal laws should lead to universal structures. A concise expression of the cognition proposal would assimilate the working of a mathematician's mind to a more or less acrobatic exercise in common sense and Descartes, with his "rules for the direction of the mind," would thus appear as a precursor of the cognition thesis.

The deductive interpretation of quantum mechanics sheds some new light on this avenue. We saw in chapter 11 how the fundamental laws of quantum mechanics turn into the familiar rules of classical physics under the effect of macroscopic decoherence. The exact form of classical physics, including Newtonian dynamics, is not the main point in the present case, but rather its practical and philosophical consequences. It was shown in chapter 13 how the prolegomena of common sense arise beyond decoherence, with causality, the locality of objects, the exclusion of incompatible properties for a unique object, and, last but not least, the validity of standard logic relating the phenomena. All these features are not, of course, restricted to the physical world and they extend with minor variations to the world of life. They were perceived, gathered, and utilized by our ancestors and they were certainly essential for the origin and the development of language.

It is clear that we agree with the cognition approach, which we used in chapter 2 when introducing the idea of classical reality. It also provides a sensible explanation for the origin of human reason, which is an obvious requirement for developing mathematics. The only criticism we can make of Changeux is that his analysis is incomplete. He does not reflect on the regularities of nature and how they can generate common sense, considering these two ideas as obvious though his interpretation of mathematics completely relies on them. Physism fills these two gaps. It brings into much more importance the effect of the laws and it can answer the two basic questions that were completely left aside by the cognition approach, namely, those of the fecundity and consistency of mathematics. It brings attention to the subtlety of the laws, which even great biologists sometimes tend to ignore, because they work in a macroscopic world and the laws they know best look much like mechanisms. This subtlety of the laws, which seem at variance with common sense in the quantum world and generate nonetheless the rules of common sense through decoherence, this subtlety was universally ignored in all the previous approaches to the philosophy of mathematics. It is in contrast a strong point of physism.

## Quasiempiricism

Quasiempiricism is another approach sharing many common points with physism but remaining incomplete. We may borrow from Hersh the idea that many concepts of mathematics originated in an inspiration from the surrounding world, after some abstraction, and most historians of mathematics agree on that point. We may also easily agree with the quasiempiricists when they interpret the practice of mathematics and stress its analogies with the practice of natural sciences. Physism goes even further to say that there is no difference between mathematics and theoretical physics, except for the origin of the problems and the final assessment of a physical theory by experiment.

When one comes to questions of method, the analysis of the method in mathematical research by Lakatos is widely acknowledged and his elucidation of the to-and-fro movement between conjectures, lemmas, and proofs is certainly more convincing than the older conceptions. He recognized, roughly speaking, that the process of mathematical discovery does not start uniquely from the conjecture of a theorem, but it comes with a series of other auxiliary conjectures. They may bear on the different possible formulations of a supposed theorem, its potential extent, and its conditions of validity; they often yield also a tentative pattern of proof, which can be broken up into a series of hypothetical lemmas. The investigation of the lemmas and of their mutual connections can end up, in the best cases, with the achievement of a valid proof and the assertion of the theorem, but it can also reveal unsuspected obstacles. These obstacles can mark impossibilities, or they may open further opportunities, from which the whole process will start anew. A great theorem is often by now the result of a collective enterprise, involving many interacting actors and in which collaboration, competition, and critique enter equally. That is in any case Lakatos' thesis on method.

There are a few weak points, however, in quasiempiricism, which we now consider.

### Quasiempiricism and the Invention of Mathematics

Lakatos' analysis of the process of proof insists on the social aspects of research. He therefore draws the benefits of a sociological approach, with a better understanding of the scientific community and the organization of research. One may wonder, however, whether these aspects teach us something essential about math. They tell us nothing about its nature, and some important aspects of the scientific method are practically left out. Lakatos' approach relies explicitly on Popper's, which is definitely noncognitive. It deals therefore rightly with the construction of a proof and the collective recognition of its validity, but the conceptual step in the four-step method (as explained in the Appendix to chapter 14) is completely ignored, although it exists in mathematics as much as in natural sciences. When this essential component of the scientific method is trivially reduced to a conjecture, one may wonder whether science is not deprived of its main contact with "mathematics itself."

Few quasiempiricists—except Thom—paid much attention to the question, "How does one conceive, or invent, mathematics?" or any science for that matter, if one assumes with quasiempiricism that the empirical method is universal. None gave a hint that would help us, for instance, to understand the role of the great architects of mathematics, like Gauss, Riemann, Poincaré, and Hilbert.

## Fallible Mathematics

Another dubious aspect of quasiempiricism is again more concerned with its dialectic than its doctrine. It consists in the repeatedly asserted fallibility of mathematics, which is supposed to corroborate its empirical character, because either men are fallible or bugs cannot always be suppressed in a software program. Lakatos illustrated this fallibility by giving long lists of errors by famous authors on important questions, but this procedure is a sophism since all the mistakes had been corrected meanwhile and he learned of them from their correctors.

The main worry one might have about this point of view is not concerned, however, with the truth of mathematics. One could be sarcastic and say that, on the contrary, the insistence of quasiempiricists on the social aspects of science did not go far enough and they ignored the possible consequences of their own dialectic on the nonscientific society. Some quasiempiricists were probably unaware of the growing influence of postmodern relativism, but Lakatos was definitely in the "postmodern" trend and his dialectic must be considered with caution. A common denominator of postmodernism is to reduce science to the status of an ordinary discourse, with no privilege over other types of discourse and no special right to claim privileged access to truth. The influence of this skeptic philosophy is presently increasing in occidental society, perhaps because it does not require any special study and nevertheless allows anyone to pass judgement on any knowledge. The main response, lacking in quasiempiricism and emphasized by physism, is of course that science is not primarily a discourse, but a dialogue with reality, which is our master.

In this dialogue, mathematics is in some sense the weak point of science in its search for objectivity, if only because of its uncertain ontological status outside physism or Platonism. This is why the thesis of mathematical fallibility must be submitted to a sharp critique and not easily accepted. As a matter of fact, the arguments supporting fallibility are often very contrived, sometimes almost grotesque; and many of them appear as sophistic quibbles, when they do not simply express the regret of the lost paradise of truth that Hilbert had promised. The importance of correction processes, which is so great in computer science and is also stressed, fortunately, by quasiempiricism, is certainly relevant for a better understanding of mathematics.

## Falsifiers and Conclusion

Lakatos himself stressed the last weakness of quasiempiricism, which consists in the lack of falsifiers ascertaining mathematical truth. We discussed that point already in the previous chapter and we can now end with the relation between quasiempiricism and physism. Quasiempiricism sometimes went too far in its assertions, perhaps because it would look too weak otherwise to stand as a philosophy of mathematics, the philosophy whose lack was stressed by Goodman, and which may be provided by physism. They do belong to a common trend, renewing—in my opinion—Poincaré's approach after a long separation of mathematics from natural science and bringing them together again, and I am tempted to consider quasiempiricism as a forerunner of physism. Their differences are otherwise plain enough and I will not insist on them, except for the essential one: physism provides an explanation for the main problem of mathematical philosophy, which is the source of its consistency and its fecundity, without an appeal to Platonism as Putnam had assumed.

## GELL-MANN AND PENROSE

There is presently a trend tending to reconsider the meaning of the laws of nature, and it has, of course, a repercussion on

221

physism. Murray Gell-Mann (1995) and Roger Penrose (1989, 1994, 1997) have written influential books showing, in my opinion, a close connection with the thesis in the present book.

## The Meaning of Quantum Mechanics

Gell-Mann is certainly the greatest physicist alive. His contributions to the modern theory of particles are fundamental and he was also a leader, in collaboration with James Hartle, of the great renewal in the interpretation of quantum mechanics, along with Robert Griffiths, Wojciech Zurek, Hans-Dieter Zeh, and others. In his book *The Quark and the Jaguar,* he did not develop the consequences of his vision of physics for mathematics, but physism is so much inspired by considerations of quantum laws that I will depart for a moment from my reverence for Pascal's dictum, "the I is contemptible," and become more personal, for the sole purpose of making clear how physism came about.

My own contributions to the collective work on quantum mechanics included the recovery of standard logic from consistent histories, the correspondence between quantum and classical physics through microlocal analysis, and, more recently, a rather general theory of decoherence. This coverage gave me the occasion to write down the first deductive measurement theory, which led me to reconsider some older philosophical problems that had been raised by the Copenhagen interpretation and its aftermath. I once asked myself, almost unwillingly: What has now changed in this stuff? Asking questions is the devil's trap for would-be philosophers, and I was caught. Who can resist a beautiful question?

Questions are like bacteria: they multiply by division and one of the progeny of mine was: Why does quantum mechanics need an interpretation? The answer was pretty obvious: because the only access to it is through mathematics. Another round of division then yielded: Why is mathematics unavoidable in fundamental physics? I pass over a few rounds extending for years, which ended in this answer: Because mathematics, at the fundamental level, is a face of the laws of nature. I then read some books on the philosophy of mathematics, which were either already familiar, new, or somewhere in between,[1] and I found no compelling objection to the proposal. The friends I talked to, all of them physicists, agreed more or less, and I thought that this answer was in the air, a fleeting and still partly unspoken idea, which had only to be written down after some investigation.

I had, however, underestimated the demands of philosophy. The first version of my endeavors, in French, was widely ignored or frowned upon. I had indeed made the mistake of centering the book on ontology, probably under the influence of the present fashion for realism. But philosophers want only good ontology respecting the rules of the art; mathematicians dislike it—particularly when it comes from a physicist—and most physicists despise it. Another division round therefore yielded the question: Why do we speak endlessly of physical reality, to conclude only that we don't yet understand it?, with the addition: But we know about the laws; why not simply analyze them and see whether an uncompromising investigation confirms the assumption of physism or disproves it? It is always almost impossible to make sure that one did not follow a presupposed inclination when proceeding to a philosophical investigation—particularly when one is not familiar with this exercise—but I believe I finally became convinced by objective and almost obvious evidence. I had then only to write this book

---

[1] This mention of "in between" corresponds to a rather surprising experience. A colleague had recommended to me a book by the philosopher André Darbon on the philosophy of mathematics and I had been looking for it in some libraries with no success. Imagine my surprise when I discovered it accidentally on a remote bookshelf at home and, when opening it, I saw its margins covered by my own handwriting of almost fifty years ago. I mention this story because I am aware that I surely felt many forgotten influences during this long period, that now enter unacknowledged into the proposal of physism.

a second and a third time, then in a foreign though beloved language, and there we are.

I began this section by referring to Gell-Mann and, finally, I didn't say anything of his thought during that long digression. This is because I think that physism is in no way a work of mine, but only the work of a transcriber who felt something in the air, or something in the present *episteme* as learned people would say.[2] A few people contributed to create that spirit of the time, particularly Feynman and Gell-Mann, and Murray was the one who influenced me most directly. As the saying goes, he shares no direct responsibility for the present work, however, and I leave the reader to find the hints of physism as I believe I noticed them in his book.

### *Penrose and Physism*

Roger Penrose wrote a series of three books in which many brilliant ideas occurred, in which I find some connections with physism. As far as the philosophy of mathematics is concerned, he claimed himself a Platonist, but his thesis concerning the invention, or discovery, of mathematics is more original. He considered the vision of math as a nonalgorithmic process and he gave inspiring arguments in favor of this idea. It means that creating mathematics is a matter of strategy and inspiration, which cannot be mimicked by a computer. The human brain is irreplaceable and the invention of a conjecture is the most decisive event in this construction.

I believe that this idea is worth considering from two different standpoints. Penrose himself is close to physism when he wants to link the nonalgorithmic character of math to the existence of a similar process in the brain, related to the basic laws. This process would be due to the occurrence of decisive quantum effects in some little components of the

[2] *Episteme* was introduced by Michel Foucault (1973) to denote the spirit of the time in science, with its inspiration and limitations.

nerve cells (the microtubules), but that is not enough, since the quantum laws are algorithmic.[3] The nonalgorithmic process would be due to quantum reduction, which Penrose envisions as due to an effect of the gravitational type ensuring the permanence of a unique space-time. This is not, in my opinion, a very promising line of research and I prefer considering the basic idea in a different light: the creation of a significant part of science (a superb part, in Penrose's words), or the discovery of a perspective in its architecture, would be a higher example of pattern recognition. Of course, to speak of "recognition" in the case of mathematics may have a Platonist undertone, but it makes sense in physism, since the method is supposed the same in every science.

One may then interpret Penrose's proposal as asserting that pattern recognition, at least in the very important case of math, is not algorithmic. Penrose mentions as a consequence that a manmade machine cannot create mathematics. Nobody can tell, but my point in the present case was twofold. It was first to mention the relation of the discovery of math and natural laws as a case of pattern recognition. I had no occasion previously to raise that point and this is a convenient one; it means that the present approach to pattern recognition in the framework of artificial intelligence can be useful in practice, but probably too narrow in principle. The second point was to say that physism (which I attribute rightly or wrongly to Penrose's episteme) should not be identified with a mechanistic vision. It was already mentioned in the previous chapter that it is open to the discovery of deeper laws than the ones we know, and the fact that the present ones are understood as algorithmic is not necessarily definitive.

---

[3] Penrose's proposal is rather doubtful on that point. He did not appreciate, nor even fully master, the consequences of decoherence in an aqueous medium. Since then, decoherence has been experimentally confirmed and straightforward calculations have disproved his assumption (Hepp 1999).

## Formalism

One might consider formalism as the achievement of quasiempiricism, when it succeeds so well that it reaches its ultimate limit. Zermelo's argument, for instance, when he justified the axiom of choice because of its fecundity was basically (quasi)-empirical. It seems also that the mathematical community tends nowadays to attribute an increasing value to fecundity among the important virtues of a work, as shown, for instance, in some recent awards of the Fields medals. I wish to indicate anyway why formalism remains essential.

### Formalism on Computers

After Gödel and Turing, formalism was reduced essentially to the writing and checking of mathematics on a computer. With Chaitin, Gödel's theorem turned into a case of noncomputability. One may expect that this tendency will become more and more important, conceptually with future development of computer science and pragmatically with the possibility of really practicing formalism and not just invoking it as a matter of principle. A significant change in the spirit of mathematics occurred when Appel, Haken, and Koch, in 1976, used a computer to complete the proof of a solution of the four-color problem. It provoked a kind of earthquake in the mathematical community, but its appreciation is now more sober, and the community has realized that this event marked in fact a return to the grand tradition of formalism.

Some opponents of the profane use of a computer in the sacred enclosure of Great Proofs argued that a computer program is always susceptible to bugs. Some quasiempiricists, like Tymoczko, argued on the contrary that human beings are still more vulnerable to errors, but we already discussed this side issue of fallibility. The attainment of an axiomatic form always marks, in any case, the achievement of a successful quasiempirical investigation and, with the advent of computers, formalism will probably be considered more and more as the expectation, the future, and the achievement of a quasiempiricist and perhaps a physicalist approach, which nourishes it.[4]

## The Meaning of Formalism

The existence and the success of formalism, not only in logic and mathematics, but also in the foundations of physics, is a deep indication of the transcendence and the unity of the laws. Let us add some comments on that.

Euclid's books stand out as the historical paradigm of formalism. It may be worth recalling that the pattern of this unique work probably originated, after a dialectical and didactical process, as the best recipe for anticipating the questions and the objections of an ever-discussing Greek audience. It resulted anyway in an impressive, abstract, and beautiful work of art, influencing hundreds of generations. It cannot be decided, however, whether formalism also initially satisfied a human desire for order. Some ancient philosophers, who took mathematics as an argument, believed that this desire is imprinted in our minds by the sanctity of Nature. In a more modern mood, one may also think that this imprint was put in our brains by Nature herself throughout the history of mankind. In any case, something deep is at stake, and although formalism was certainly brought forth through quasiempirical labor, its success is a wonder of nature.

The longing of the nineteenth-century mathematicians to regain a stable foundation for their science had essentially the same origin, in an unsatisfied eagerness for order and harmony. Perhaps some founders, like Frege or Hilbert, expected too much, but that does not diminish their achievements.

---

[4] I use "physicalist" here as the adjective associated with the noun "physism."

They turned Cantor's paradise, as Hilbert called the new mathematics, from a thick and obscure forest into a luminous garden.

Something similar happened in physics with its foundation on axiomatic principles, although the process was smoother. Physics knew an earlier drama, with the advent of quanta when startling experimental results required unprecedented concepts, so that the need for order was more easily satisfied when its time came. The relevant point is not therefore in that case how Einstein could place the theory of relativity on firm foundations, or how Dirac, Von Neumann, and later physicists made those of quantum physics and particle physics apparent, but it resides in a philosophical assessment of the resulting principles.

The foundations of physics now stand on definite principles, which can be written down in a few pages—most of them in a mathematical language. They are abstract, of course, but their number is amazingly small. The achievement of such a tight consistency is certainly not simply the outcome of a work of art, a human masterpiece of economy and sobriety, but something whose germ must belong to the laws themselves. It reveals an intrinsic order, which is depicted in human science by an axiomatic form. That seems to be true of the laws of nature and of mathematics as well, as the expression not of a vague analogy between them but of a community of essence.

Some mathematicians tend to underestimate nowadays the axiomatic approach, because the present period is rather one of efflorescence. The recent achievements of particle physics arose, on the contrary, from a quest for a deeper substructure of the laws, which became in the last few decades a still more ambitious search for pure consistency involving physics and mathematics altogether. Physism shares this belief in an undivided science, beyond transient epistemes. The axiomatic form of the laws of science should be considered as a great philosophical discovery of the scientific enterprise and nothing like it reveals so clearly the overall consistency of knowledge.

## Intuitionism

Intuitionism, when reduced to the idea of founding mathematics exclusively on the integers, is an exception in the present case, because of its incompatibility with physism. One may acknowledge of course that these natural numbers are "natural," in so far as they seem reasonable to the human mind. They are probably imprinted genetically in our brains since, as shown by Piaget, they are among the earliest manifestations of understanding in the behavior of infants. One might even say that the genetic code is the most beautiful embodiment of the integers in base 4, but this is not enough for asserting intuitionism.

The concept of integers cannot be primary, or at least not uniquely primary, from the standpoint of physism. It rests on the existence of finite sets (of people, toys, or animals) which are certainly obvious in our surroundings, but fundamentally second to the fundamental laws! Macroscopic objects, neatly separated in spite of their similarity and able to enter in a finite set, exist only beyond the edge of decoherence. Below that level, the fundamental laws rely as far as we know on a nonconstructive basis including Hilbert spaces, quantum observables, gauge fields, and so on, assuming the axiom of choice, which is the pet aversion of intuitionism. Elementary particles cannot replace the elements of a finite set as a basis for a theory of integers, because they are undistinguishable.

In a Bourbakian exposition of the laws of nature starting from some axiom one for mathematics, the integers could enter early and—why not?—be among the first listed, but the idea that they constitute a universal basis for a consistent philosophy of every significant part of mathematics is—at least as far as I can see—excluded by physism.

## Physism as a Synthesis

There was often an impression among mathematicians that every school in the philosophy of mathematics holds a part of the truth but is incomplete. The various schools were deemed incompatible and were always opposed in the philosophical discourse, so much so that Hersh could recently summarize the situation by saying that "each one of them fetishizes one aspect of mathematics and insists that that one limited aspect *is* mathematics" (Hersh 1997) There was nonetheless a hope that this conflict is not fate. There should be a way to bring unity among them and, as Goodman said, "[This expected] philosophy of mathematics would be only one chapter in a larger philosophy of science. That philosophy would make clear in what sense there is only one objective world and how it is that the objects studied by the mathematician, many of which are not related with physical reality, can nevertheless be seen as parts of that world."

We have shown in this chapter how the most sensible ideas of cognition sciences, quasiempiricism, and formalism can be integrated into physism, and how they complete each other. This is a completely new situation in the philosophy of mathematics. The most important aspect of this comparison is probably the parallel existing between the axiomatic structure of mathematical formalism and the existence of basic principles in physical laws. The unity thus attained in the philosophy of mathematics relies ultimately on the consistency of science itself, which must be considered as a whole.

# Physism: A Discussion

**W**hen I was writing the last version of this book, an expert told me the rules of the art, when somebody proposes a supposedly new philosophical thesis. He or she should first explain the thesis, then compare it with existing ones, and, finally, answer anticipated objections. The first two steps have now been duly fulfilled and we arrive at the last one. I thought initially that it would be a dull academic exercise and I was ready to leave it out. But, when I began, I discovered that it could give rise to years of research, leading perhaps to unexplored perspectives. I did not try, however, to pursue these shadows and they will be mentioned briefly, particularly at the end of this chapter.

## THE OBJECTIONS AGAINST PHYSISM

When I tried to think of possible objections to physism, the first ones I thought of were concerned with its interest. Does it after all bring anything new or valuable? The main interest of an idea is to generate other ideas, and one might therefore question physism from that standpoint. The main schools in the philosophy of mathematics, logicism, formalism, and intuitionism, for instance, raised interesting questions inside mathematics itself, which led sometimes to interesting theorems; what about physism? It does not immediately suggest a clear-cut program of investigation, at least at first sight. One may also feel some disappointment when observing that physism does not apparently shed any new light on the problem of realism. Perhaps, I thought, these questions will clear up later, but my program today is to work in the garden and finish that chapter. These comments can wait and they are not, after all, objections against physism and its veracity. If somebody raises them, I'll give the usual answer, namely, that a new point of view may help to enlarge one's vision, and a reasonable answer is always interesting when the question itself is. That will do.

## The Objection from Bohmian Mechanics

The only obvious objection I could anticipate came from thinking of the physicists and philosophers of science who are engaged actively in the program of "Bohmian mechanics." This theory, which was proposed by David Bohm, is a tentative version of quantum mechanics in which one has, as usual in the nonrelativistic case, a wave function obeying the Schrödinger equation and depending on the position of a number of particles under consideration (Bohm and Hiley 1993). These particles possess moreover their own "realistic" degrees of freedom: they are supposed to be located at "real" well-defined positions with well-defined velocities, and their real motion is governed by a very clever combination of ordinary classical forces and of quantum forces depending on the wave function. A nice feature of this construction is the agreement it obtains between the usual quantum probabilities—which are given by the square of the wave function—and the peculiar statistical distribution of the real positions. This agreement is obtained as follows: One assumes that the real positions of the particles are random at an initial time, with a probability distribution (as meant in the classical sense) coinciding with the quantum distribution resulting from the wave function. The dynamics of the particle is chosen so cleverly that the two distributions—quantum and classical—remain identical at later times. Some sensible arguments have also been proposed that explain the initial coincidence rather naturally (Dürr et al. 1992).

This theory (which is by the way much better known in philosophy departments than in physics departments) is highly appreciated for the elegant solution it provides for the problems of realism. As mentioned in chapter 7, it says that elementary particles are genuine particles with well-defined positions and velocity, the only correction to classical reality being their peculiar dynamics, which depends on a wave function. If Bohm's theory were proved experimentally, it is clear that much of what we said about the characters of the laws would be revised, the Hilbert space formalism would become an insignificant mathematical superstructure, all our discourse on the role of potentialities in the laws would boil down to ordinary probability calculus, and, in a nutshell, physism would be empty.

### The Main Objection against Bohmian Mechanics

The Bohmian theory has serious difficulties itself, in so far as it has no sensible relativistic version (as already mentioned in chapter 8) and, in particular, it deals very awkwardly with the creation and annihilation of particles. Particle physicists observe that relativistic quantum field theory—with creation and annihilation of relativistic particles—was formulated before 1930, one to three years after the discovery of nonrelativistic quantum theory. In the case of Bohmian mechanics, which was proposed in the early 1950s, no satisfactory relativistic version is yet known half a century later. There is, for instance, no satisfactory theory of optics in this framework, and one does not even know whether the electromagnetic field or the photons are the real objects in that case. When some special relativistic feature of particles is included in the Bohmian framework, moreover, its clumsiness looks awful when compared to the elegance of standard relativistic quantum field theory.

The standards of some philosophically motivated people and those of physicists are completely at variance in that case. Particle physicists are deeply impressed by the elegance and precision, the detailed agreement with experiments over a tremendously large range of magnitude, and seventy years of conceptual progress and discoveries that are due to relativistic quantum field theory. Remarkably enough, no particle physicist is working on Bohm's theory, as far as I know: the conceptual gap and the difference in criteria are too wide. The many philosophers who wish so eagerly to believe in this theory are not aware of the importance of this relativistic difficulty, and Bohm's idea, initially so brilliant, widens the gap between physics and philosophy, whereas it might have filled it up rapidly, if it had been right.

When anticipating therefore the Bohmian objection, my answer will be that the theory should first be shown to be consistent with special relativity, with the presently immense collection of well-explained experimental data, and also as internally consistent as standard quantum field theory. Then the failure of physism would be, in any case, a negligible event as compared with the resulting revolution.

### *Why There Is No Relativistic Bohmian Mechanics\**

Everything hinges therefore on the lack of a sensible relativistic version of Bohmian mechanics. Why, then, is it so difficult to get it? The answer to which most physicists incline is that there is certainly no such theory. But is there a simple and demonstrative reason for that? There is none, of course—otherwise there would be no point in the present discussion. I think, however, that there is at least an explanation, which turns out to be related to the critique we made of intuitionism in the last chapter. This is why I thought it worth mentioning here, in spite of some technical aspects.

We saw earlier (in chapter 5, the section "On Identical Particles") that one of the basic principles of quantum mechanics asserts that identical particles are indistinguishable. Dirac's argument (which we gave in that section of chapter 5) has a continuation, which is less well known in spite of the importance that Sam Treiman put on it in his book *The Odd Quantum*. He showed beyond doubt that the existence of a quantum field depends on the fact that the particles described by the field are strictly indistinguishable! It seems therefore that the difficulty of Bohm's theory with relativity is due to two conflicting requirements, namely, (1) quantum fields are necessary in relativistic physics, and that implies undistinguishable particles; (2) the existence of "real" particle positions in Bohm's theory implies a possibility of distinguishing them.

Point 1 is substantiated by the enormous amount of data resulting from quantum electrodynamics and from the standard model. Point 2 is perhaps less obvious and I will add a few comments on it. If a particle had a "real" position $y$ in a definite reference frame at a given initial time, one could use $y$ as the label of the particle. But a difficulty appears when one tries to build up a quantum field description of particle creation, because one must specify the point where the particle is "really" created, according to the Bohmian form of realism. The mathematics of quantum fields—which I cannot develop here—imply then that, in place of a local quantum field, say $\psi(x)$, one must be dealing with an indexed field $\psi(y, x)$ (notice that this is a field and not a wave function). This means that one must introduce a different field for every particle, since the *real* position labels are different. The resulting field theory then involves an infinite number of different fields for a unique type of particle and, although such exotic fields have been studied in other instances,[1] their properties are very different from those of the standard quantum fields. In the case of photons and the electromagnetic field, for instance, one can account for the behavior of a weak classical field, but

---

[1] For instance, when quantum field theories associated with Regge poles were studied.

it seems impossible to recover the coherence properties of laser light.

## PHYSISM AND THE CARTESIAN PROGRAM

The second point in this discussion is concerned with the relation of physism to the Cartesian program. It could imply an objection of a higher order, more properly philosophical, which I am not sure to perceive and certainly not to master, but which is undoubtedly important. The Cartesian program (according to Heidegger's expression), or the Cartesian philosophy of science, was described in a strong form by Husserl (1970) in *The Crisis of European Sciences and Transcendental Phenomenology*, a collection of his work on this topic including also transcendental phenomenology, which he proposed as an alternative to the Cartesian approach. I believe by the way that his enterprise of phenomenology is incompatible with the existence of quantum mechanics, but that is not the point. The question I want to address is whether physism implies Husserl's formulation of the Cartesian philosophy and the accompanying criticisms.

I will be brief, because this topic could immerse us completely, and, furthermore, I prefer not to expose the extent of my ignorance. One may summarize the Cartesian program, according to Husserl, as asserting that every feature of reality can be described in a mathematical form. Its relation to physism seems a vast question, which I intend to restrict to a discussion of the problem of quantum reduction, which was already mentioned in chapter 13 (the section "On the Uniqueness of Classical Reality"). If one assumes—with Roger Penrose and some other authors in that section—that there exists a physical reduction process ensuring a universal uniqueness of classical reality, then it would seem that physism implies the validity of the Cartesian program. If, on the other hand, Gell-Mann and Griffiths are right in saying that the quantum laws cannot determine a way out of the probabilistic framework, then the

Cartesian program fails: there are events in this world that cannot be completely accounted for by means of mathematics. The outcome of a quantum measurement stands in that case as a counterexample to a complete mathematical description of everything in reality.

The philosophical alternative, which I proposed and which was also described in chapter 13, is explicitly at variance with the strong form of the Cartesian program. If there were an ontological chasm between reality and the web of the laws enclosing it, if the laws accounted for everything in reality, except for its essence, which is its uniqueness, then there would be a gap in the Cartesian program. There would be a difference in essence between the mathematical laws of nature and nature itself, or reality, assuring a creative role of time as in Bergson's vision. This is, however, a discussion using "if" and it would be pointless to try to say more than that presently. The conclusion is simply that physism does not necessarily imply the Cartesian program.

## How Mathematics Is Made

The question of "how mathematics is made" was already discussed in chapter 3 from a practical standpoint, and we briefly encountered the corpus and the tricks of the trade of mathematics; some remarks were also added in chapter 14 when the main philosophies of mathematics were described. We may agree essentially with the quasiempirical description of the practice of mathematics by a philosopher like Lakatos, or by mathematicians like Thom or Hersh. Axiomatism and formalism appear in that case as successful achievements of quasiempiricism, although these various philosophical frameworks do not really bear on the everyday work of the mathematicians.

The actual question is not, however, about the practice and method of mathematics, but why they succeed. I believe that physism goes much farther than the earlier approaches on that matter, and it renovates particularly the fundamental issues of fecundity and consistency. It also extends the philosophical methodology for discussing various features of mathematics, because it allows a comparison with similar aspects of theoretical physics and a reference to the laws of nature. I came upon these questions only at the last moment, for purely rhetorical reasons, when I was trying to fill up the gap that occurred at the beginning of this chapter, namely, to answer the question of the program of physism, or the next round of investigation it suggests. It looks so vast now that I will mention a few examples. The main suggestion appearing through these examples is that every "why?" question concerning the success of a mathematical method points toward a corresponding aspect of the laws of nature, and it emphasizes a little more the shortcomings of our collective thinking on the meaning of these laws.

### Elementary Geometry

Consider, for instance, the beginnings of geometry, particularly when Egyptian scribes computed the area of a triangle as in figure 1.1. The idea of straight lines, together with parallel and orthogonal ones, was then the first empirical concept, which led to so much development later. The question, " Why did these concepts succeed?" may look pointless, but its relation to the underlying laws of nature is easily pointed out. Suppose for instance that the ratio $G/c^2$ of the gravitational constant to the square of the light velocity were much bigger, then geometry, according to the laws of general relativity, would not look Euclidean. It would depend strongly upon neighboring objects and, assuming that the idea of using geodesics in place of straight lines had occurred, a parallel to (i.e., a curve at a fixed distance from) a straight line would not have been a straight line, and several perpendiculars could be drawn from a point to a straight line. One may notice also, from the standpoint of differential geometry, that a Euclidean space is related to its tangent space by homeomorphism, and that undoubtedly made the discovery of calculus much easier.

The wonderful simplicity of the Euclidean space, which lies ultimately in its physical simplicity, explains a lot about the origin of mathematics. This paragon of spaces has a wonderful group of invariance—a physical invariance, of course—an invariance that is particularly propitious to abstraction. Serious geometry began with the idea of identical (or equal) figures, particularly triangles, and this idea was inspired by invariance under a displacement. When, much later, Sophus Lie was inspired by the corresponding group, it is impossible to decide whether the meaning of the group follows from the existence of the laws, or whether the laws cannot be expressed without the notion of invariance and the existence of an underlying group. There is no weak junction where these mathematical concepts and physical laws can be dissociated.

## Logic

Elementary logic would stand as axiom number zero in an imaginary Bourbakian account of science, but on the other hand we saw in chapter 11 that the conditions for empirical common sense emerge only a long way after the basic laws. Consideration of the historical origin of logic, in relation to these laws, is not therefore an empty question. We shall make only a few indicative remarks, which are not meant to question standard logic and some related concepts, but only their blatancy.

The notion of self-identity, which leads to the basic logical equivalence $A = A$, came certainly from the observation of permanent objects, with unchanging shape and some other conserved characters, such as color, under a displacement.[2]

---

[2] Jean Piaget observed that the notion of permanent objects appears rather late during the intellectual development of children (at the age of "cuckoo, there it is!").

But it would not hold if $G/c^2$ were much bigger, as shown by the complexity of the notion of a solid object in general relativity. By the way, a displacement would have to be very slow in special relativity (with a much smaller value of $c$), because otherwise the object would change color (and also shape).

The notion of distinct objects is also a long-shot consequence of quantum mechanics. When the structure of atoms was discovered, people were impressed by their emptiness: just a few pointlike particles in an ocean of vacuum. Why do separated objects exist in these conditions? Why don't they penetrate each other and mix together? If this happened, the intuitive notion of well-defined objects entering as elements in the constitution of a set would be untenable. The answer turned out to be Pauli's principle on the symmetry of a wave function under the exchange of identical particles, and it certainly looks far fetched for people who take elementary set theory as something obvious! By the way, we mentioned already how, if we could see indistinguishable particles, they would also pull the rug from under that theory.

More generally, the recognition of a common pattern in different objects or different beings, which is the starting point of abstraction, obviously rests on the existence of underlying laws. Without the laws, at a fundamental level allowing the properties of matter, and without the extraordinarily subtle decoherence effect generating classical reality, mathematics would be unthinkable and, as a pure product of thought, impossible.

### Time and the Complex Numbers*

I wish now to add a few remarks about a topic that may look pedantic, but which is nonetheless fascinating: I mean the strange connection linking time with complex numbers. It came out first when Euler realized that the function $\exp(-2\pi i \nu t)$ represents a point turning on a circle at a frequency $\nu$. The description of harmonic motions had been made previously with the help of the trigonometric functions $\sin(2\pi\nu t)$ and $\cos(2\pi\nu t)$, but the calculations were curiously simpler when the complex exponential was used. The explanation came later, when it was realized that this function is an irreducible representation of the group of time translations (for instance, the group of operations resulting from a change in the origin of time). So far, so good; but then it was discovered that causality—in the weak sense that an outgoing signal cannot precede the ingoing signal producing it—implied that many physical quantities (for instance, a refractive index, an electrical conductivity, or a magnetic susceptibility) are functions of the complex variable $\nu$, when the frequency is extended to complex values. Well, it was apparently a curiosity, but the Cauchy integral formula for functions of a complex variable implied that, generally, absorption can be derived from transmission, and conversely. For instance, one can compute the absorption of light from the knowledge of its real refractive index, and conversely. There are many more consequences, which I will not describe, in statistical mechanics and many other fields of physics.

But things become still more surprising when one comes to quantum mechanics. Every quantity or operation of a physical nature is represented in this theory by linear transformations on the wave functions or vectors in Hilbert space, except one, namely, time inversion: the operation $T$ consisting in replacing the time $t$ by $-t$. This operation is antilinear, which means that it transforms every complex quantity into its complex conjugate! A beautiful theorem, resulting from subtle properties of the functions of several complex variables, asserts then that locality in quantum field theory—essentially the fact that every interaction takes place at a definite point in space-time—implies a universal invariance of physics under the product $CPT$ of the charge conjugation $C$ (replacing every particle by the corresponding antiparticle), the inversion $P$ (parity) of the space axes, and the time inversion $T$. OK, this is again an interesting curiosity; but wait! We come now to the

241

field theories of the standard model. They are so much constrained by gauge symmetries that no complex parameters, no violation of the time inversion $T$ can occur, at least as long as the number of quark-lepton families is less than three. But there are three families. Nobody knows why, at least from the standpoint of consistency, but a consequence is the possibility of $T$ violation. "Who cares," you might say, we cannot change the direction of time! That's true, but if $T$ can be violated, the same is true for the product $CP$ and that can be observed (it has been). It would again look like a curiosity, if it did not have a tremendous consequence: the $CP$-violating interactions are most probably responsible for the fact that matter (and not antimatter) survived from the great furnace at the origin of the universe. Then God saw that complex numbers are good.

## Conclusion

I did not try to exhibit the relation between the most elementary and the most evolved mathematical notions and the corresponding features of the fundamental laws. It would certainly be a very difficult exercise, involving a wide culture in mathematics, physics, history, and philosophy. If the future of physism is to generate a research program, however, this one would certainly be in it. Another facet of the same enterprise would be a more systematic philosophical investigation of the characters of fundamental laws, which is surprisingly absent—as far as I know—from the present literature.[3]

[3] See, however, Bas Van Fraassen (1990).

# Ontology

I avoided, until the last moment, the question of ontology. The main reason was a personal conviction, which increased while working on the project of this book, that it was possible to draw reasonable philosophical conclusions from an analysis of the present knowledge on mathematics and the laws of physics, without committing oneself to extreme assumptions. The second reason was that some previous work I did, in *Quantum Philosophy* and in the first French version of my reflections on mathematics, was much more spoiled than improved by a foolhardy reliance on ontology. I learned then dearly how right Kant was in his warning.

There are two reasons, however, for a return to this slippery ground. The first one is to say at least a few words about realism, which I have avoided so carefully, if only to say more plainly why I did so. The second one has to do with Platonism. Its compatibility with several other versions of the philosophy of mathematics was mentioned earlier, and that applies also to physism. I therefore thought that the feedback of physism on Platonism was worth a few comments. I insist though on the fact that the present last chapter is not a part and parcel of the rest of this book, but an addendum, a comment for interested outsiders, at best a possible opening, which I myself find tempting although I do not wish to force it on my fellow readers.

## About Realism

We saw in some detail the chasm between the quantum world and classical reality. We also saw how decoherence, which is a quantum effect, relates them. That did not help us much, however, in understanding reality itself, and we could talk only of its laws. We might assume anyway that reality is unique and shows two faces, existential and essential, and that would bring us toward d'Espagnat's idea of a "veiled reality," with perhaps the advantage of a clearer identification of the "veil." We might also insist on the chasm between the two faces, which would become essential if reduction (discussed briefly in chapter 11, "The Multiplicity of Quantum Frameworks and the Uniqueness of Classical Reality") were left finally with a "philosophical" answer asserting that reality and its laws are so deeply separated that nothing more can be said. It would look like Kant's distinction between phenomena and noumena, where, however, we would know a great deal about noumena, in view of our approximate knowledge of the laws; the phenomena, on the other hand, would be plunged into a wider extension, which we may call quantum reality or anything similar. One might call this position *dualism*. Its main advantage would be an economy of thought: we concentrate on what we can know, the laws and classical reality, and we recognize that there is something else, of the same nature as classical reality though only knowable by means of the laws.

One may mention in this respect that the idea of keeping apart the laws and physical reality came from the late Aristotelian tradition, particularly Posidonius, St. Augustine, and Simplicius. Hipparcus introduced it originally in Alexandrine astronomy, in view of the alternative explanations of planetary motions by means of epicycles and eccentric trajectories of the planets. When Augustine spoke, for instance, of the motion of Venus, he said: "Astronomers have tried to express this motion in various ways. But their assumptions are not necessarily true since the appearances one sees in heavenly bodies might perhaps be saved by some other form of motion yet unknown to man." The famous motto according to which science "saves appearances" reappeared many times in the history of ideas, and Pierre Duhem used it as the title of one of his books. Laws and reality were kept apart by the "philosophers" as they were called in antiquity (for instance the neo-Platonists with Plotinus, Porphyrus, and Jamblicus), whereas the laws were held as a true and complete knowledge of reality by the "physicists" (such as Democritus and Epicurus) who believed in the possibility of an exhaustive knowledge.

Dualism brings new features into this controversy. In its perspective, the laws and reality are both objective, but their duality makes realism nontrivial. Physism, by drawing attention to the intimate relation of the laws and mathematics, could not avoid including logic in the same circle. Dualism is then led to raise the question of the relevance of logic when investigating the relation of reality and its laws. What can be logically said of that "relation"? If logic belongs to the laws, however, it cannot reach outside its domain and account for a so-called relation—just a word—that would involve reality itself. So, do we know only appearances when we learn about the laws? Maybe, but, presently, there is nothing else we can *know*.

I had no occasion to recount in this book the flow of speculations on reality glutting the literature. The most promising ones, in my opinion, suppose that another change in the characters of the laws will occur at the level of Planck's scale, and that could refresh realism while negating dualism. Maybe, but a wise attitude seems to be to wait and see, and as far as I am concerned, I would say that I am certainly not a realist, in the sense of the word one can hear in some academic circles discussing the philosophy of physics, because I cannot understand what it means. It looks to me only like bad—and of course useless—ontology.

## On Platonism

Mathematical Platonism is a paradigm of reification: concepts that are sublimated into a specific reality. Roughly speaking, it said traditionally that the totality of mathematics belongs to an intrinsic reality, its own, which is distinct from the empirical one. Whereas constructivism would see mathematics as a grand product of our imagination, Platonism reifies it into something existing by itself. We saw variants of Platonism, in which, for instance, Gödel attributed an ontological existence to the concepts of set theory, whereas Thom extended this existence to the whole of mathematics. Most Platonists conceive this reality as so tightly linked with mathematics that it is confined in some sense to that science. The human mind explores a remote world, with toil and hardship, sometimes with joy, and it brings back little nuggets of knowledge or large visions, composing our mathematics.

Physism and, in some sense, dualism are compatible with Platonism, although it should be stressed that they do not require it or imply it. The ideal reality of a dualistic Platonism cannot be restricted to the sole content of mathematics, but it could stand at the nodal point of the physical laws, which we called on some occasion *Logos*. The ontological *Logos* would stand outside space and time. Its architecture— i.e., its own laws—would be considered as the ordinary Platonic "reality" behind mathematics, so that there would after all be some agreement between the old and new forms of Platonism. This architecture could stand on primary pillars that would be represented by the axioms of mathematics and the basic principles of natural sciences. Another fundamental feature of *Logos* would be its general organization under logic. Since the ontological *Logos* is supposed to be intrinsically distinct from the physical reality, it does not need to share its uniqueness and actuality.

## *The Extent of Logos*

Two questions cannot be avoided in this framework. The first one is concerned with the relation of the dualistic character of physism to classical philosophy. Many clues might certainly be exploited if inspiration were taken from Bergson and Spinoza, to mention just two. I know just enough philosophy to guess so, but I am too ignorant to dig deeper. The second question is suggested by the religious meaning of the word *"Logos,"* although the use of another name would not have avoided it. I introduced it, of course because of its meaning in Greek philosophy, but one knows that the paradigm of realism—from which mathematical Platonism derives through Descartes and the lessons of philosophy from his Jesuit teachers—was patterned in medieval universities on the content of God's mind. The question "What is the extent of *Logos* in a Platonist version of physism?" will then certainly be asked by some people expecting a metaphysical or religious continuation of the story.

I think anyway that any statement of mine, either in that direction or against it, would wreck the spirit of this book, which was conceived as an *analysis* of knowledge with speculations kept at a minimum level, and I will add nothing more about ontology.

## "Physism" versus "Physicalism"

A side remark finds its proper place here, however. I am aware that, when developing an analytic approach, I was led to favor physics as the main body of science with relevance to the philosophy of science, whereas physism was supposed to stand on a consideration of nature, which is much wider than just the consideration of physics. There are two reasons for this slip of thought. The main one is that physics is still unique as a highly formalized science so that its relation to mathematics is indeed a privileged one. Another more personal reason is due to my own experience of that science and the limitations of my knowledge of others.

Physics has often been a source of inspiration for mathematicians, but there are other significant sources. Life sciences and human sciences, for instance, played a great role in the development of probability calculus. The present progress of biology raises a host of new mathematical problems, which are only in their beginnings. To mention only one, it becomes clear that one is now facing what might be called the "manifold" problem. If that name were adopted, it would not refer, of course, to geometric manifolds in the usual sense, but to the structure or the dynamics of systems involving many variables possessing somewhat different statuses, for instance, systems having many different time scales and/or many organized levels. The obvious example is the case of a living being with its organs, its cells, the components of the cells, and the molecules in them.

When trying to tackle these problems, one suspects that a new level of mathematical research is needed, which might start a new era. Historically, after all, mathematics began with systems involving only a few variables, or a few features. The twentieth century has shown us how much can be learned about infinite systems, but "manifold systems" have been little investigated. A deeper understanding of life and—why not—of human sciences might very well require them. Physics is certainly not the right place for such investigations, because physical systems are most often too uniform (although Kenneth Wilson's theory of second-order phase transitions might be considered as an example of a "manifold theory," its success was based on the uniformity of the various levels when one splits, for instance, a system of many spins into small aggregates of spins, these aggregates joining into larger ones, and so on).

One might surmise that this very young kind of mathematics will at some time become essential if the mathematical sciences extend their range toward living systems and social systems. But they do not presently, so that an analytic approach to the corresponding future philosophy of mathematics is premature, to say the least. Anyway, I wished to mention these far-fetched perspectives as a reminder of the transitory status of the relation between mathematics and the laws of nature as they are known at any specific time. It goes almost without saying that if such speculations were some time to be true, the question of ontology would be deeply changed, just as it might be if the present trend in research toward a wider unification of physical laws (through noncommutative geometry, string theory, or something else) gives rise to new hints.

## The Sense of It All

"The meaning of it all" is the title of a book collecting some inspiring lectures by Richard Feynman, published posthumously. The publishers did not make clear whether Feynman himself wanted this title, but it surely nicely portrays the content. You may notice that the heading of this ultimate section is slightly different, "meaning" being replaced by "sense." The nuance I wish to introduce distinguishes "meaning," as something of the order of the intellect, from "sense," taken as closer to sentiment. When I say "sense" in the present case, I intend to suggest a "sense for us," us human beings, who are confronted with the vastness of the universe and the impassibility of its laws.

When I tried to express what the laws and their form afford as a sense for us, I could find no better way than reproducing what I wrote earlier on that point in *Quantum Philosophy* (I apologize for quoting myself, with the benefit of Arturo Sangalli's excellent translation). I had previously recalled Einstein's saying: "The conviction that the world is governed by rational rules and can be apprehended by reason belongs to the domain of religion. I cannot conceive a true scientist lacking this profound belief. The situation may be expressed through an image: science without religion is lame, religion

249

without science is blind." Then I added the following (with a few minor changes)

It seems to me that we must distinguish here between two words. Einstein's reflection acquires its full significance if "religion" is understood in the sense of "sacred." The latter captures a concept that Mircea Eliade introduces in the foreword of his *Histoire des croyances et des idées religieuses* (*A History of Religious Beliefs and Ideas*) as follows:

"It is difficult to imagine how the human mind could operate without the conviction that there is something irreducibly *real*[1] in the world; and it is impossible to imagine how consciousness could appear without conferring a *meaning* on human impulses and experiences. Consciousness of a real and meaningful world is intimately connected with the discovery of the sacred. By experiencing the sacred, the human mind has grasped the difference between real, powerful, rich and meaningful things and others not possessing these attributes, that is, the chaotic and dangerous flow of events, their haphazard and meaningless occurrence and disappearance, and not a mere stage in the development of consciousness. . . . In short, the "sacred" is an element of the structure of consciousness, and not a mere stage in the development of this consciousness.

If we compare this conception of the sacred to the definition in a dictionary (the French dictionary *Robert* in this case) we notice a similarity: something is sacred that "merits an absolute respect, which may be considered as an absolute value." This is quite different from another common denotation: that which is sacred "belongs to a separate domain, forbidden and inviolable (as opposed to what is profane), and inspires a sentiment of religious reverence." This second meaning is accompanied by references to words such as "saint" and "taboo." It is preferable

---

[1] The author's italics.

to exclude this latter sense, since it establishes a duality clearly absent from Einstein's idea.

Mircea Eliade actually defines sacred twice in the text cited above, and his two definitions are different: he considers it first as something powerful and significant in itself, and later as a way to experiment this power by a particular disposition of consciousness that he regards as a structure of the latter. We are not in a position to decide whether the sacred is a disposition of consciousness (in which case it would be generated by the harmony of reality) or as a cultural inclination; it would be enough to admit that it is a state of consciousness that many, if not all, of us know under some form or another. The important point is to grant that the sacred is a disposition experienced by the individual, and that it therefore establishes a relationship between the world and human behavior or—why not?— between a philosophy of knowledge and humanity.

Hence, it is the first quality that Eliade attributes to the sacred that is for us the most important: the quality of being powerful, rich, and meaningful. We may also observe that some of Eliade's reservations are unnecessary. When he talks about "the chaotic and dangerous flow of events, their haphazard and meaningless occurrence and disappearance," he seems to assume that this domain, repulsive to the sacred, may belong to some primal reality independent of any form of order. Now, we know that such a disorder is only apparent: ill-fated circumstances or a tragic accident may appear fearsome or fatal, but they are nevertheless governed by a higher order, one closer to the laws. The flow of things may be dangerous and loaded with risks for the individual, the group, or even the species, but there is nothing chaotic in its essence, even though it remains complex and unpredictable. The appearance and disappearance of things may seem to occur at random, but they are never senseless. To sum up: the way we see it, the sacred is everywhere in the universe and nothing is completely profane. Profanity is but an illusion of our own ignorance, the slumber of the mind, or the madness of our false ideas.

And to conclude, not only this chapter but also the book itself, I will put my trust in one of the great discoveries of modern brain science, which is that there can be no thinking without a cortege of emotions and sentiment (Damasio 1994, 1993, 2003). This is why I do not care whether there exist one or two realities, or more, I do not confide my beliefs to leaky words and fuzzy ontological dreams, and I do not mind that our knowledge of the laws is transitory. What "makes sense for me," and I hope for you, is that the laws—and their inseparable world of mathematics—should be hailed as sacred.

# Bibliography

Aharonov, Y., and L. Vaidman (1991). Complete description of a quantum system at a given time. *J. Phys.* A 24: 2315–2328.

Alley, C. O., O. Jakubowicz, C. A. Steggerda, and W. C. Wickes (1983). Page 136 in *Proceedings of the International Symposium on the Foundations of Quantum Mechanics*, S. Kamefuchi, ed. Tokyo: Physical Society of Japan.

Aspect, A., P. Grangier, and G. Roger (1981). Experimental tests of realistic local theories via Bell's theorem. *Phys. Rev. Lett.*, 47: 460–463.

Baire, R., E. Borel, J. Hadamard, and H. Lebesgue (1905). Cinq lettres sur la théorie des ensembles. *Bull. Soc. Math. France* 33: 261–273.

Bell, J. S. (1964). On the Einstein Podolsky Rosen paradox. *Physics* 1: 195–200.

———(1987). *Speakable and unspeakable in quantum mechanics*. Cambridge: Cambridge University Press.

Benacerraf, P., and H. Putnam (1983). *Philosophy of mathematics: Selected readings*. Cambridge: Cambridge University Press.

Bohm, D. (1951). *Quantum theory*. New York: Prentice-Hall.

———and B. J. Hiley (1993). *The undivided universe*. New York: Routledge.

Bohr, N. (1958). *Atomic physics and human knowledge*. New York: Wiley.

Bourbaki, N. (1960). *Éléments d'histoire des mathématiques*. Paris: Hermann.

Bouveresse, J. (1988). *La force de la règle: Wittgenstein, les mathématiques et le monde reel*. Paris: Minuit.

Brune, M., E. Hagley, J. Dreter, X. Maître, A. Maali, C. Wunderlich, J. M. Raimond, and S. Haroche (1996). Observing the progressive decoherence of the "meter" in a quantum measurement. *Phys. Rev. Lett.* 77: 4887–4890.

Cartwright, N. (1983). *How the laws of physics lie.* London: Routledge.

Chaitin, G. (1982). Gödel's theorem and information. *Int. J. Theor. Phys.* 21: 941–954; reprinted in Tymoczko (1998), pp. 287–311.

Changeux, J-P., and A. Connes (1995). *Conversations on mind, matter and mathematics.* Princeton, N.J.: Princeton University Press.

Connes, A. (1994). *Noncommutative geometry.* New York: Academic.

———(2000). Non-commutative geometry, year 2000, arXiv:math. QA/0011193 (23 Nov).

Cornwell, J., ed. (1995). *Nature's imagination: The frontiers of scientific vision* (preface by F. J. Dyson). Oxford: Oxford University Press.

Damasio, A. R. (1994). *Descartes' error: Emotion, reason, and the human brain.* New York: Putnam.

———(1999). *Feeling of what happens: Body and emotion in the making of consciousness.* New York: Harcourt Brace.

———(2003). *Looking for Spinoza: Joy, sorrow, and the feeling brain.* Orlando, Fla.: Harcourt.

Daneri, A., A. Loinger, and G. M. Prosperi (1962). Quantum theory of measurement and ergodicity conditions. *Nucl. Phys.* 33: 297–319.

Darrigol, O. (1992). *From c-numbers to q-numbers: The classical analogy in the history of quantum mechanics.* Berkeley: University of California Press.

Davis, P.-J., and R. Hersh (1986). *Descartes' dream: The world according to mathematics.* New York: Harcourt Brace Jovanovich.

Dieudonné, J. (1972). *Eléments d'analyse,* 9 vols. Paris: Gauthier-Villars.

———(1978). *Abrégé d'histoire des mathématiques (1700–1900),* 2 vols. Paris: Hermann.

———(1992). *The music of reason.* Berlin: Springer.

Dirac, P. A. M. (1930). *Quantum mechanics,* Oxford: Oxford University Press.

———(1963). The evolution of the physicist's picture of nature. *Sci. Am.* 206 (5): 45–53.

Dixmier, J. (1964). *Les C\*-algèbres et leurs représentations*. Paris: Gauthier-Villars.

Duhem, P. (1989). In *La théorie physique: Son objet, sa structure*. P. Brouzeng, ed. Paris: Vrin.

Dürr, R., S. Goldstein, and N. Zanghi. (1992). Quantum equilibrium and the origin of absolute uncertainty. *J. Stat. Phys.* 67: 843–907.

Einstein, A., B. Podolsky, and N. Rosen (1935). Can quantum-mechanical description of physical reality be considered complete? *Phys. Rev.* 47: 777–780.

Elitzur, A. C., and L. Vaidmann (1993). *Found. Phys.* 23: 987.

D'Espagnat, B. (1995). *Veiled reality: An analysis of present-day quantum mechanical concepts*. New York: Addison-Wesley.

———(2002). *Traité de physique et de philosophie*. Paris: Fayard.

Feyerabend, P. (1975). *Against method*. London: Verso Books.

Feynman, R. P. (1965). *The character of physical laws*. Cambridge, Mass.: MIT Press.

———(1998). *The meaning of it all*. Reading, Mass.: Perseus Books.

Feynman, R. P., R. B., Leighton, and M. Sands (1965). *The Feynman lectures on physics*. Reading, Mass.: Addison-Wesley.

Foucault, M. (1966). *Les mots et les choses; Archéologie du savoir*, Paris: Gallimard. English translation, *The order of things; An archeology of the human sciences*. New York: Vintage Books, 1973.

Gell-Mann, M. (1994). *The quark and the jaguar: Adventures in the simple and the complex*. New York: Freeman.

Gell-Mann, M., and J. B. Hartle (1991). Quantum mechanics in the light of quantum cosmology. In *Complexity, entropy and the physics of information*. W. H. Zurek, ed. Redwood City, Calif.: Addison-Wesley.

———(1993). *Phys. Rev.* D 47: 3345.

Giulini, D., E. Joos, C. Kiefer, J. Kupsch, O. Stamatescu, and H. D. Zeh (1996). *Decoherence and the appearance of a classical world in quantum theory*. Berlin: Springer.

Gödel, K. (1995). *Collected works*. Vol. 3, *Unpublished essays and lectures*. S. Feffermann et al., eds. Oxford: Oxford University Press.

Goodman, N. D. (1979). Mathematics as an objective science. *Am. Math. Monthly* 86: 540–551; reprinted in Tymoczko (1998), pp. 79–94.

Greenberger, D. M., M. A. Horne, A. Shimony, and A. Zeilinger, Bell's theorem without inequalities. *Am. J. Phys.* 58: 1131–1143.

Greene, B. (1997). *The elegant universe: Superstrings, hidden dimensions and the quest for the ultimate theory.* New York: Vintage Books.

Griffiths, R. (2002). *Consistent quantum mechanics.* Cambridge: Cambridge University Press.

Hardy, L. (1992). Quantum mechanics, local realistic theories, and Lorentz-invariant realistic theories. *Phys. Rev. Lett.* 68: 2981–2984.

Hawking, S. (1988). *A brief history of time.* London: Bentham.

Henderson, G. W. (2002). Book review. *Mathematical Intelligencer* 24 (1): 75.

Hepp, K. (1999). "Toward the demolition of a computational quantum brain." pp. 92–104 in *Lecture Notes in Physics.* Vol. 517, *Quantum future,* Ph. Blanchard and A. Jadszyk, eds. Berlin: Springer.

Hersh, R. (1979). Some proposals for reviving the philosophy of mathematics. *Adv. Math.* 31: 31–50; reprinted in Tymoczko (1998), pp. 9–28.

———(1997). *What is mathematics, really?* Oxford: Oxford University Press.

Hilbert, D. (1992). In *Natur und Mathematisches Erkennen: Vorlesungen, gehalthen 1919–1920 in Göttingen.* D. E. Rowe, ed. Boston: Birkhäuser.

Hoddeson, L., L. Brown, M. Riordan, and M. Dresden (1997). *The rise of the standard model: A history of particle physics from 1964 to 1979.* Cambridge: Cambridge University Press.

Hörmander, L. (1985). *The analysis of linear partial differential equations.* Berlin: Springer.

Husserl, E. (1970). *The crisis of European sciences and transcendental phenomenology.* D. Carr. trans. Evanston, Ill.: Northwestern University Press.

Jaffe, A., and F. Quinn (1994). Theoretical mathematics: Towards a cultural synthesis of mathematics and theoretical physics. *Bull. Am. Math. Soc.* 29: 1–13.

Jammer, M. (1966). *The conceptual development of quantum mechanics.* New York: McGraw-Hill.

Kneale, W., and M. Kneale (1978). *The development of logic.* Oxford: Clarendon.

Kuhn, T. (1962). *The structure of scientific revolutions.* Chicago: University of Chicago Press.

Lakatos, I. (1976). In *Proofs and refutations.* J. Worrall and G. Currie, eds. Cambridge: Cambridge University Press.

———A renaissance of empiricism in the recent philosophy of mathematics? In *Philosophical papers,* J. Worrall and G. Currie, ed. Cambridge: Cambridge University Press; reprinted in Tymoczko (1998), pp. 29–48.

Lakoff, G., and E. Nunez (2000). *Where mathematics comes from: How the embodied mind brings mathematics into being.* New York: Basic Books.

Logothetis, N. (2000). La vision: une fenêtre sur la conscience. *Pour la Science* 268: 80.

Omnès, R. (1999a). *Quantum philosophy.* Princeton, N.J.: Princeton University Press.

———(1999b). *Understanding quantum mechanics.* Princeton, N.J.: Princeton University Press.

———(2000). *L'espion d'ici.* Paris: Flammarion.

———(2002). Decoherence, irreversibility, and selection by decoherence of exclusive quantum states with definite probabilities. *Phys. Rev.* A 65: 052119 [1–18].

Penrose, R. (1989). *The emperor's new mind: Concerning computers, mind, and the laws of physics.* Oxford: Oxford University Press.

———(1994). *Shadows of the mind: An approach to the missing science of consciousness.* Oxford: Oxford University Press.

———(1997). *The large, the small and the human mind.* Cambridge: Cambridge University Press.

Polya, G. (1945). *How to solve it.* Princeton, N.J.: Princeton University Press.

———(1954a). *Induction and analogy in mathematics.* Princeton, N.J.: Princeton University Press.

———(1954b). *Patterns of plausible inference.* Princeton, N.J.: Princeton University Press.

Popper, K. (1959). *The logic of scientific discovery*. London: Hutchinson.

Putnam, H. (1975). What is mathematical truth? In *Philosophical papers*, 2 vols. Cambridge: Cambridge University Press. Reprinted in Tymoczko (1998), pp. 49–66.

———(1988). *Representation and reality*. Cambridge, Mass.: MIT Press.

Queneau, R. (2001). *Zazie in the metro*. Barbara Wright, trans. New York: Penguin Books.

Schweber, S. (1994). *QED and the men who made it: Dyson, Feynman, Schwinger, and Tomonaga*. Princeton, N.J.: Princeton University Press.

Steiner, M. (1998). *The applicability of mathematics as a philosophical problem*. Cambridge, Mass: Harvard University Press.

Tymoczko, T. (1998). *New directions in the philosophy of mathematics: An anthology*. Princeton, N.J.: Princeton University Press.

Van Fraassen, B. (1990). *Laws and symmetry*. New York: Oxford University Press.

Van Heijenoort, J. (1967). *From Frege to Gödel: A source book in mathematical logic*. Cambridge, Mass.: Harvard University Press,.

Van Kampen, N. G. (1954). *Physica* 20: 603.

Von Fritz, K. (1945). The discovery of incommensurability by Hippasus of Metapontium. *Ann. Math.* 46 (2): 242–264.

Von Neumann, J. (1932). *Mathematische Grundlagen der Quantenmechanik*. Berlin: Springer.

Wang, H. (1987). *Reflections on Kurt Gödel*. Cambridge, Mass.: MIT Press.

Weinberg, S. (1993). *Dreams of a final theory*. London: Vintage Books.

———(1996). *The quantum theory of fields*. 3 vols. Cambridge: Cambridge University Press.

Wheeler, J. A., and W. H. Zurek. (1983). *Quantum theory and measurement*. Princeton, N.J.: Princeton University Press.

Wigner, E. P. (1970). *Symmetries and reflections; Scientific essays of Eugene P. Wigner*. Cambridge, Mass.: MIT Press.

Wittgenstein, L. (1956). *Remarks on the foundations of mathematics*. Oxford: Blackwell.

Zeh, H. D. (1970). On the interpretation of measurement in quantum theory. *Found. Phys.* 1: 69–76; Reprinted in Wheeler and Zurek (1983), pp. 342–349.

———(2001). *The physical basis of the direction of time.* Berlin: Springer.

Zurek, W. H. (1991). Decoherence and the transition from quantum to classical. *Phys. Today* 44 (10): 36.

———(2003). Decoherence, einselection, and the quantum origin of the classical. *Rev. Mod. Phys.* 75, 715.

# Index

Abel, Niels, 33
abstraction, viii, 10–11, 12, 166–67
Aristotle, vii, 23–24; and abstraction, 10–11
Archimedes, 26
axiom of choice, 42, 181; in physics, 91–94
axiomatism, 35, 183

Balian, Roger, 120n
Bell, John, 46
Benacerraf, P., 179
Bennett, Charles, 214
Bergson, Henri, 155, 163, 237
Birkhoff, George, 113
Bohm, David, 102–3, 172, 232–35
Bohr, Niels, 46. *See also* complementarity; correspondence principle
Born, Max, 161–62
Bourbaki, Nicolas, xi, 7, 12, 26, 37, 182, 183
Broglie, Louis de, 82
Brouwer, Jan, xiv, 186. *See also* intuitionism
Burali-Forti, Cesare, 41

C*-algebras, xii. *See also* Gelfand, Alexander
Cantor, Georg, 38, 40
Caratheodory, Constantin, 207
Cartan, Elie, 203
Cartesian program, 66, 236–37
Cartier, Pierre, 90n
causality, 170
Chaitin, G., 44, 186, 187, 212, 226

Changeux, Jean-Pierre, 12n, 190, 216, 217
choice. *See* axiom of choice
classical science, definition, 21
cognition: and mathematics, 12–14, 190; and physism, 216–18
Cohen, Paul, 42, 211
complementarity, xiv, 131
complex numbers, 29; and time, 240–42
complexity, 187
Connes, Alain, 151, 189, 212n, 216
consciousness, 17
consistency, x, 183
consistent histories. *See* Griffiths, Robert B., histories of
constructivism, 188
correspondence principle, 46
crisis: in mathematics, ix, 39–44; in physics, 44–47

Dali, Salvador, 19
Damasio, Antonio R., 252
decision problem, 184, 187
decoherence, xv, 114–25, 164, 168, 217, 244; and classical behavior, 123–25; theory of, 120–21; observation of, 122–23
Dedekind, Richard, 183, 217
Descartes, René, viii. *See also* Cartesian program
determinism, 168–70
Dieudonné, Jean, 26, 30n, 182
Diosi, C., 172

Dirac, Paul, 152, 206–7, 228; equation of, 148–49; and observable quantities, 67
dualism, 244
Dürr, D., 172

Egorov theorem, 169
Einstein, Albert, 148, 151, 168, 195, 228, 249; and Einstein-Podolsky-Rosen paradox, 46, 123
Eliade, Mircea, 250
environment, 117–18
episteme, 224n
Espagnat, Bernard d', 141n, 244
Euclid, 25–26, 227
Eudox, 10
Everett, Hugh, 172

falsification, x, 193, 197, 221
fecundity: in mathematics, xi, 31–33, 181–82
Fermi, Enrico, 205
Feynman, Richard, 128n, 195, 249; and Feynman histories, 51–62, 157, 162, 204, 211; and Feynman rules, 99–101
formalism, viii, 182–85; and physism, 226–28
Foucault, Michel, 224n
Fourier, Joseph: and Fourier series, 33–34; and physism, 27
four-color problem, 226
four-stage method. See method
Fréchet, Maurice, 203
Frege, Gottlob, 40, 180, 181, 228
Frenkel, A., 172

Galois, Evariste, 32
Gardner, Martin, 214
Gelfand, Alexander, xii, 93
Gell-Mann, Murray, 14, 105, 172, 204, 216, 221–24, 236
Ghirardi, G. C., 172
Gibbs, Josiah, 207

Gödel, Kurt, 189, 226, 246; incompleteness theorem of, 36, 44, 185, 187, 212
Goldstein, S., 172
Goodman, Nicholas D., 191, 194, 195, 221, 230
Griffiths, Robert B., 105, 172, 222, 236; histories of, 113, 128–34, 169
Grothendiek, A., 203

halting problem, 187
Hankel, Hermann, 38
Hartle, James B., 105, 172, 216, 222
Heidegger, Martin, 67, 155, 163, 236
Heisenberg, Werner, 76–80. See also uncertainty relations
Hepp, Klaus, 225n
Hermite, Charles, 37
Hersh, Ruben, viii, 165, 167, 189, 191, 193, 194, 218, 230, 237
Hilbert, David, xiv, 40, 228; and Hilbert space, 84–85; on physical theories, 106–7, 200; program of, 42–44
Hörmander, Lars, 125, 132
Hume, David, 13–14
Hurwitz, Adolf, 204
Husserl, Edmund, 46, 66, 236

identical particles, 73–75
interference: of the macroscopic type, 110–12; pattern of, 61, 116n
intuitionism, 182, 185–86; vs. physism, 229
inverse problems, 33, 208

Kant, Emmanuel: and categories, xiv, 20; schemata of, ix; and time, 16. See also Pauli, Wolfgang
Karolihazy, F., 172
Klein, Felix, 202
Kuhn, Thomas, 147

Lakatos, Imre, 190–93, 218, 220, 237
Lamb shift, 158, 160

Laplace, Pierre Simon de, 161
Laws of nature: characters of, 141–63; consistency in, 146–52; creative character of, 153–54; existence of, 142–46; invariant form of, 158–60; and potentialities, 160–62; vs. time, 155
Leibniz, Gottfried Wilhelm, viii
Lichnerowicz, André, 205
Locke, John, 13–14
logic: elements of, 22–26, 239; nonstandard form of, 113; and quantum mechanics, 130–32
logicism, 179–81

mathematics: the making of, 30–37, 207–8, 237–40; and reality, 37–39
"manifold" problem, 248
measurement. *See* quantum measurement
method, 195–98
metamathematics, 35, 184
microlocal analysis, 124, 166, 168, 203
Minkowski, Hermann, 202
model, 36
Morette-De Witt, Cécile, 90n
noncommutative geometry, 151
nonstandard analysis, 212

observable, an, 67–68
omega number, 213
Omnès, Roland, 120n, 170, 195, 237
Onsager, Lars, 207

paradoxes: counterfactual paradoxes, 133; in quantum mechanics, 133–34
pattern recognition, 198, 225, 240
Pauli, Wolfgang, 45
perception, 15–19
Peano, Giuseppe, 40, 41, 181, 183
Pearle, P., 172
Penrose, Roger, 172, 189, 224–25, 236

Petitot, Jean, 197
Physism: and the Cartesian program, 236–37; and cognition, 216–18; definition of, xiii, 137, and formalism, 226–28; history of, 199–205; vs. intuitionism, 229; and the making of mathematics, 237–42; objections to, 231–33; and quasiempiricism, 218–21; as a synthesis of other approaches, 230; theses of, 208–9, 214–15
Piaget, Jean, 229, 239
Planck's length, 150
Plato, vii, 8–10
platonism, 185, 188–190, 246–47
Poincaré, Henri, 37, 186, 198, 221; and physism, ix, 38, 200
Polya, G., 35
Popper, Karl, x, 195
positivism, 194
projection, 108
proof, 184
propositions, in quantum mechanics, 109–10
Putnam, Hilary, 179, 189, 191, 210, 221
Pythagoras, 6

quantities, 65–85; noncommutative, 80–82
quantum electrodynamics, 96–99
quantum mechanics, 45–47. *See also* Feynman, Richard, and histories; quantum measurement; quantum electrodynamics
quantum measurement, 69–70, 171
quasiempiricism, 190–95; and physism, 218–21

realism, 138, 173, 194, 244–45
reality, xiv; classical characters of, xiv, 20–21; according to quantum mechanics, 105–25, 164–173; uniqueness of, 171–73. *See also* mathematics

renormalization, 149, 151
Richard, Jules, 41
Riemann, Bernhard, 183
Riesz, Frederic, 203
Rimini, A., 172
Robinson, Abraham, 211
Russell, Bertrand, xiv, 41, 180, 181

Sangalli, Arturo, 249
Segal, Ian, xii, 93
Schrödinger, Erwin, 82–83; and the
    cat problem, 46, 111–12; and the
    Schrödinger equation, 64
Schwarz, Laurent, 150, 203
Schweber, Sam, 80
Schwinger, Julian, 80–81
Shimony, Abner, 189
Sobolev, Sergei, 203
string theory, 151
structure, 34–36, 184

Thom, René, 189, 193, 218, 237,
    246
time and complex numbers,
    240–42
Treiman, Sam, 235
Turing, Alan, 186, 226
Tymoczko, Thomas, 191, 226

ultrafinitism, 189
uncertainty relations, 71–72

Van Fraassen, Bas, 242
Van Heijenoort, Jean, 179, 180
Vasarely, Victor, 19
virtual processes, 101–4, 162
Von Fritz, K., 7
Von Neumann, John, 84, 93, 108–13,
    228

Wang, Has, 189
wave function, 62–64
Weber, T., 172
Weinberg, S., 214
Weyl, Hermann, 124, 182, 186
Wheeler, John A., 134n, 162
Wiener, Norbert, 204
Wigner, Eugene, 124, 149, 201n
Wilson, Kenneth, 248
Wittgenstein, Ludwig, 46, 163, 180

Young interference device, 53

Zanghi, N., 172
Zeh, Hans D., 105, 119n, 222
Zermelo, Ernst, 41–42
Zurek, Wojciech, 106, 222